SUCCESSFUL BUILDING PRIMER

SUCCESSFUL BUILDING PRIMER

Robert Taylor

Structures Publishing Company
Farmington, Michigan 48024

Manufactured in the United States of America

Edited by Virginia Case

Text Design by Linda A. Zitzewitz

Current Printing (last digit)
10 9 8 7 6 5 4 3 2 1

Structures Publishing Co.
24277 Indoplex Circle
Box 1002, Farmington, Mich. 48024

ISBN 0-89999-010-X {PAPER}

Library of Congress Cataloging in Publication Data

Taylor, Robert Lowell, 1903-
Successful building primer.

Includes index.
1. House construction--Amateurs' manuals.
2. Dwellings--Remodeling--Amateurs' manuals.
I. Title.
TH4815.T39 690'.8 80-19490
ISBN 0-89999-010-X (pbk.)

INTRODUCTION

THIS BUILDING PRIMER CONTAINS BASIC AND FUNDAMENTAL INFORMATION FOR USE WHEN PLANNING AND THINKING ABOUT A NEW HOME. WHETHER YOU ARE CONSIDERING LOCATION, BUILDING, BUYING, REMODELING OR RESTORATION, YOU WILL FIND VALUABLE, EASY TO UNDERSTAND SKETCHES, TABLES, CHARTS AND DATA THAT WILL SAVE YOU TIME, CONCERN AND MONEY.

YOU WILL FIND MANY THINGS TO HELP YOU WHEN DISCUSSING YOUR BUILDING PROBLEMS WITH REALTORS, BUILDERS, CRAFTSMEN, MONEY LENDERS AND DEALERS.

EXPERIENCE HAS PROVEN THE VALUE AND IMPORTANCE OF MAKING AND KEEPING NOTES AS YOU PROCEED WITH PLANS TO BUY, BUILD OR REMODEL A HOME. FOR THIS REASON, AMPLE SPACE HAS BEEN PROVIDED FOR MAKING NOTATIONS.

THE BOOK IS NOT IN EXACT ALPHABETICAL ORDER. WE SUGGEST YOU CHECK THE COMPLETE INDEX FOR THE INFORMATION YOU NEED.

INDEX

ACREAGE CALCULATIONS-----

THE USE OF THIS TABLE WILL SHOW YOU THE APPROXIMATE ACREAGE OF VARIOUS SIZED SQUARE TRACTS OF LAND.

DIMENSIONS ONE SIDE	AREA IN SQUARE FEET	ACRES
66 FEET	4,356	1/10TH
73.8 "	5,445	1/ 8TH
85.2 "	7,260	1/ 6TH
104.4 "	10,890	1/ 4TH
120.5 "	14,520	1/ 3RD
147.6 "	21,780	1/2
180.8 "	32,670	3/4
208.7 "	43,460	1
295.2 "	87,120	2
361.5 "	130,680	3
466.7 "	217,800	5

REFER TO *PLOT SIZES* IN THIS BOOK TO FIGURE THE ACREAGE OF RECTANGULAR OR IRREGULAR SIZED PIECES OF PROPERTY.

ARCHITECT SELECTION-----

IF YOUR PROJECT REQUIRES THE SERVICES OF AN ARCHITECT, TRY TO SELECT ONE WHO SPECIALIZES IN THE TYPE OF WORK YOU HAVE IN MIND AND ONE FAMILIAR WITH LOCAL CONDITIONS.

ONE SOURCE IS THE TELEPHONE BOOK. ANOTHER IS THE AMERICAN INSTITUTE OF ARCHITECTS. YOUR LIBRARY OR CHAMBER OF COMMERCE CAN SUPPLY THE ADDRESS OF THE NEAREST CHAPTER OR THE NATIONAL HEADQUARTERS OFFICE. FRIENDS OR RELATIVES WHO HAVE USED ARCHITECTURAL SERVICES CAN USUALLY SUGGEST A RELIABLE FIRM.

Notes: ____________________

AGREEMENT-----

THIS IS A TYPICAL STANDARD AGREEMENT USUALLY USED BETWEEN A CONTRACTOR OR BUILDER AND THE OWNER OF A NEW HOME OR LARGE MODERNIZATION PROJECT. ON SMALL REMODELING JOBS A SIMPLE CONTRACT OUTLINING THE DETAILS, COST, TIME AND TYPE OF MATERIALS IS ALL THAT IS NECESSARY. ALSO, SEE *AGREEMENT TO BUY OR SELL REAL ESTATE*.

THIS AGREEMENT, made in ____________________ this __________ day of ______________, 19____,

by and between ____, hereinafter called

the Contractor, and ____, hereinafter called the Owners, WITNESSETH:

That the Contractor and Owner for the consideration hereinafter named agree as follows:

ARTICLE I. The Contractor agrees to provide all the materials as specified and to perform all the labor shown on the plans and described in the Specifications for the erection of ______________

for the owner, ____________________, ____________________________

at ____, City of ______________,

County of ______________________, State of ______________.

ARTICLE II. The Contractor agrees that the work under this contract shall be substantially completed the __________ day of ______________.

ARTICLE III. The Owner shall pay the Contractor for the performance of the contract, subject to additions and deductions provided therein, the sum of $______________ payable as follows:

____

____

____

ARTICLE IV. Final payment shall be due __________ days after completion of the work and the contract fully performed.

ARTICLE V. The Contractor and the Owner agree that the General Conditions, the Specifications and the Drawing, together with this agreement form the contract, as if hereto attached.

The Contractor and Owners for themselves, their successors, executors, administrators and assigns, hereby agree to the full performance of the convenants herein contained.

IN WITNESS WHEREOF, they have executed this agreement the day and year first above written.

______________________________ ______________________________

(Witness)

______________________________ ______________________________

(Witness)

AGREEMENT TO BUY OR SELL REAL ESTATE

SALES AGREEMENTS, CONTRACTS AND REGULATIONS COVERING THE SALES OF REAL ESTATE VARY WIDELY FROM ONE AREA TO ANOTHER AND FROM STATE TO STATE. THEY USUALLY CONTAIN ITEMS SUCH AS PRICE, TERMS, CONDITIONS COVERING OCCUPANCY, AND CLOSING DATE AS WELL AS DETAILS REGARDING FURNISHINGS, DRAPES, RUGS, ETC. AS SOON AS A CONTRACT OR AGREEMENT HAS BEEN SIGNED BY BOTH PARTIES AND PROPERLY WITNESSED, THE NECESSARY WORK TO COMPLETE THE TRANSACTION SHOULD PROCEED AT ONCE. IF THE SELLER OR OWNER HAS A CLEAR TITLE TO THE PROPERTY, HE SHOULD BE IN POSITION TO IMMEDIATELY FURNISH THE FOLLOWING TO THE OFFICE OF THE REALTOR OR LENDING INSTITUTION HANDLING THE FINANCING:

- TITLE INSURANCE POLICY (IF TITLE IS INSURED)
- LAST TAX BILL OR RECEIPT FOR LAST TAX PAYMENT
- FIRE INSURANCE POLICY AND PREMIUM RECEIPT

IF THE PROPERTY IS MORTGAGED OR UNDER LAND CONTRACT OR OTHER ARRANGEMENT, THE OWNER SHOULD PROVIDE THE ABOVE AS WELL AS THE FOLLOWING:

- LAST RECEIPT OF PAYMENT ON MORTGAGE OR OTHER ARRANGEMENT
- PAYMENT RECORD OR BOOKS OR BALANCE DUE THE LENDER, LENDING INSTITUTION OR MORTGAGE HOLDER.

THE LENDING ORGANIZATION YOU SELECT WILL CHECK ALL THESE ITEMS CAREFULLY AND COMPLETELY BEFORE MAKING A LOAN. IF NO LENDING AGENCY IS INVOLVED, THESE PRACTICES SHOULD BE FOLLOWED CAREFULLY. IF YOU RETAIN AN ATTORNEY, HE WILL CHECK THESE ITEMS BEFORE CLOSING THE TRANSACTION. SAVINGS AND LOAN ASSOCIATIONS ARE ESPECIALLY CAREFUL IN CHECKING ALL REAL ESTATE TRANSACTIONS AND USUALLY MAINTAIN COMPETENT PERSONNEL TO HANDLE THESE MATTERS. COMMERCIAL BANKS AND OTHER LENDING AGENCIES GENERALLY OFFER THE SAME SERVICE AND IN ALL CASES THE SERVICE IS COVERED BY THE PREDETERMINED LOAN FEE, OR CLOSING COSTS.

Notes: ______________________________

APPLIANCES, SPACE REQUIRED-----

TO PREVENT CROWDED CONDITIONS IN YOUR KITCHEN, UTILITY, LAUNDRY OR FAMILY ROOM, USE THIS CHART TO EASILY DETERMINE THE SPACE REQUIREMENTS FOR THE APPLIANCES YOU WILL REQUIRE. THESE ARE THE APPROXIMATE DIMENSIONS OF BUILT-IN AND FREE STANDING TYPES. *Note: All dimensions are given in inches!*

BUILT-IN TYPES	WIDTHS	DEPTHS	HEIGHTS
DISHWASHERS	24	24	34-1/2
OVENS	21 TO 30	24	26 TO 41
" (STACK-ON TYPES)	31	22 TO 24	23
RANGES	24 TO 30	24	25 TO 35
" - DRAWER TYPE WITH OVEN	30 TO 40	25	23
RANGE TOPS	21 TO 41	21	VARY
REFRIGERATORS	27 TO 36	24	51 TO 84
FREE STANDING TYPES			
DRYERS	25-1/2 TO 31	25 TO 28	36
RANGE	19 TO 36	24-1/2 TO 27-1/2	36
" EXTRA LARGE	39 TO 42	24-1/2 TO 27-1/2	36
REFRIGERATORS	24 TO 35-1/2	27-1/4 TO 30-3/4	TO 71
WASHERS	25-1/2 TO 31	25 TO 28	36

APPLIANCES, OPERATING COSTS-----

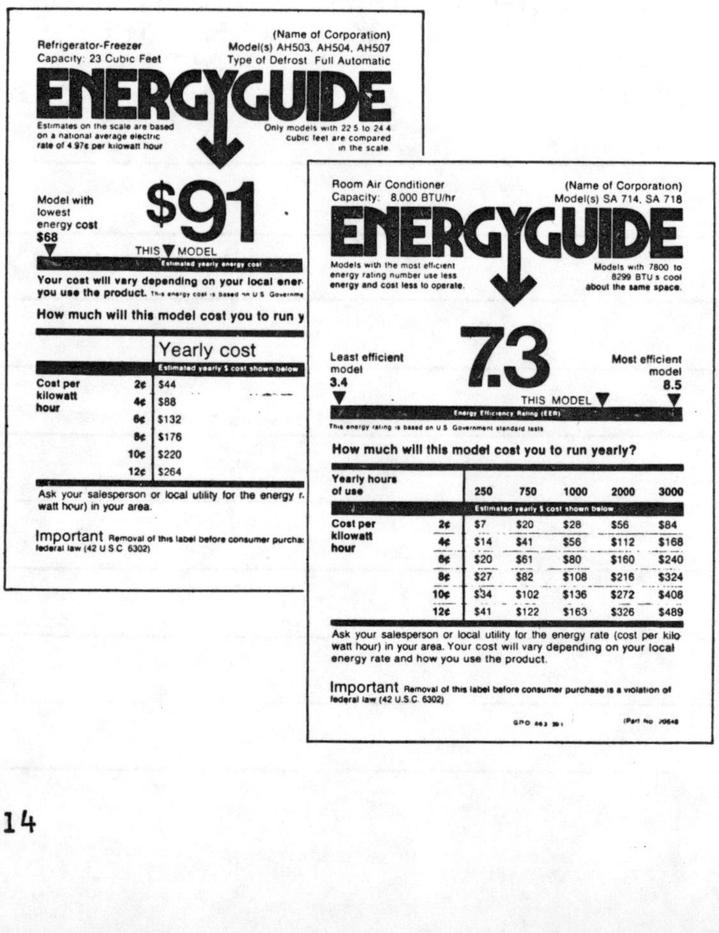

ENERGY GUIDE LABELS ARE NOW FURNISHED BY APPLIANCE MANUFACTURERS ON SEVEN TYPES OF HOUSEHOLD APPLIANCES THAT USE THE MOST ENERGY.

THE LABELS HELP SHOPPERS COMPARE THE ENERGY EFFICIENCY AND ESTIMATED OPERATING COSTS OF DIFFERENT MODELS AND BRANDS OF DISHWASHERS, REFRIGERATORS, FREEZERS, WASHERS, WATER HEATERS AND ROOM AIR CONDITIONERS.

THESE SEVEN APPLIANCES ACCOUNT FOR ABOUT 90 PERCENT OF THE ENERGY CONSUMED IN AMERICAN HOMES.

AREAS-----

THE SKETCHES BELOW SHOW SEVERAL IRREGULAR PATTERNS AND THE METHOD OF CALCULATING THE AREAS OF EACH SHAPE.

PARALLEL SIDES

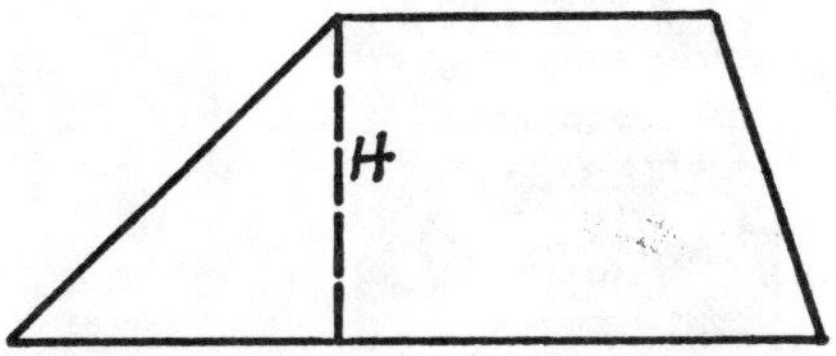

MULTIPLY 1/2 TOTAL OF SIDES BY HEIGHT (H)

PARALLELOGRAM

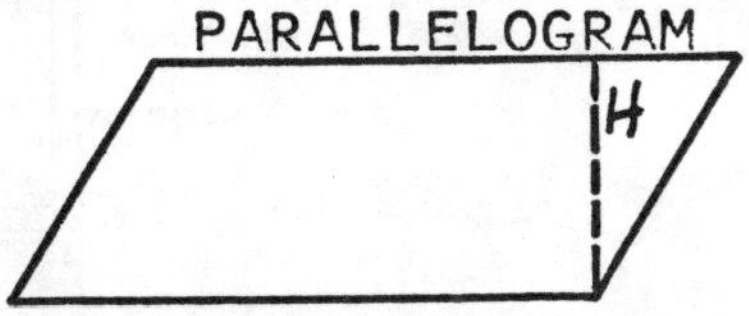

MULTIPLY LENGTH BY HEIGHT (H)

TRIANGLES

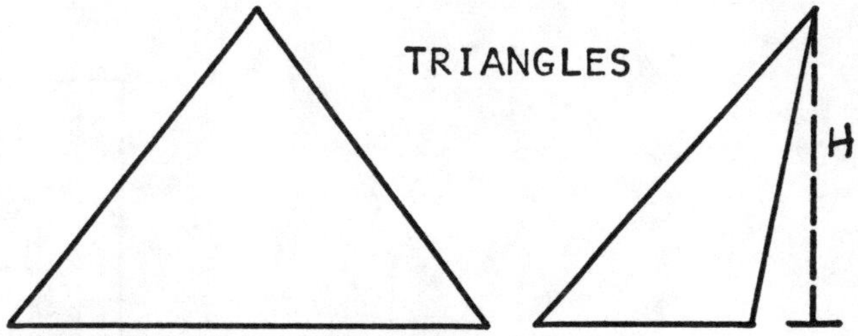

MULTIPLY WIDTH OF BASE BY 1/2 THE HEIGHT (H)

IRREGULAR SHAPES

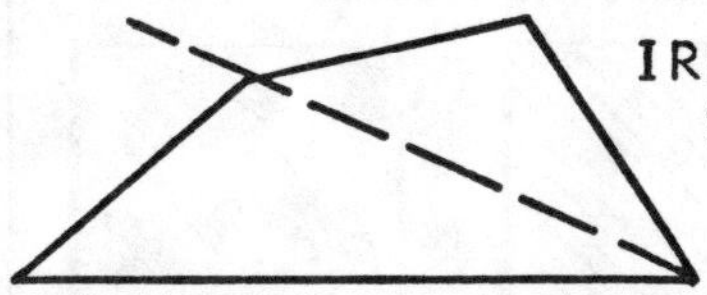

DIVIDE INTO TRIANGLES AND FIGURE EACH SEPARATELY

RECTANGLE

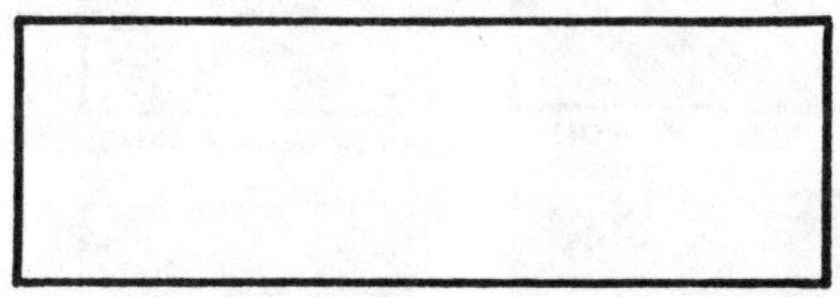

MULTIPLY LENGTH BY HEIGHT

APPAREL, SPACE REQUIRED-----

THIS CHART OF APPAREL REQUIRED SIZES WILL ASSIST YOU WHEN DETERMINING CLOSET OR STORAGE SPACE.

	WIDTH	LENGTH
MENS SUITS - COATS	22"	54"
" TROUSERS	16"	42"
" SHOES	6"	12"

	WIDTH	LENGTH
WOMENS DRESSES (REG)	14"	46"
" " (EVENING)	18"	70"
" SUITS	18"	30"
" SHOES	4"	10"

APPROXIMATE DEPTHS REQUIRED: 24" TO 30".
HEIGHTS DEPEND UPON CEILING HEIGHTS — USUALLY 90" TO 96".

Notes: ____________________

ARCHITECTURAL DETAILS-----

THESE SKETCHES OF MATERIALS AND DETAILS WILL HELP YOU UNDERSTAND YOUR HOME PLANS.

MATERIAL SYMBOLS

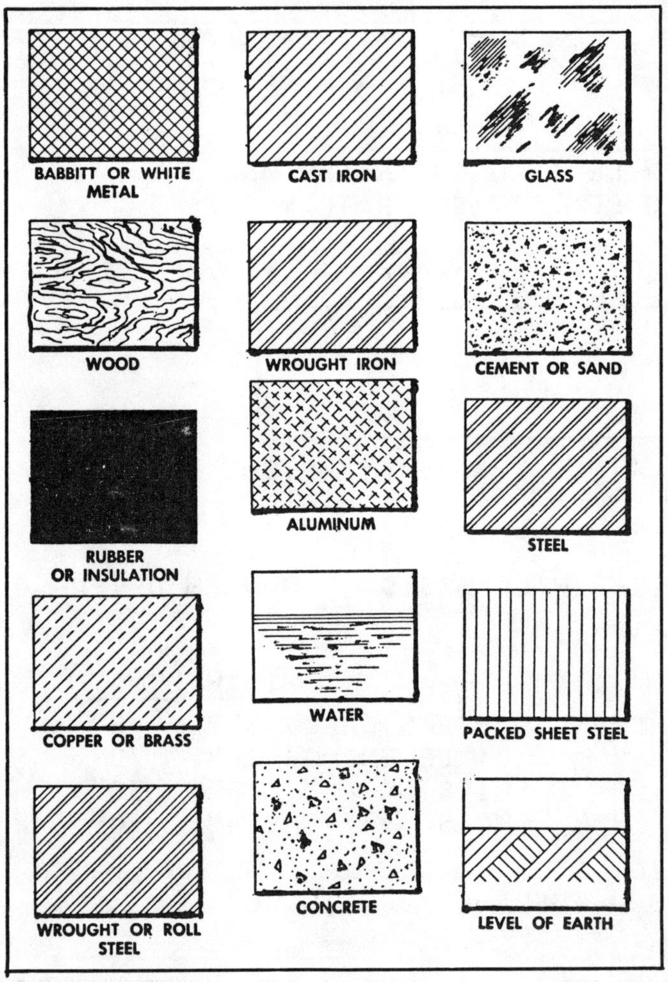

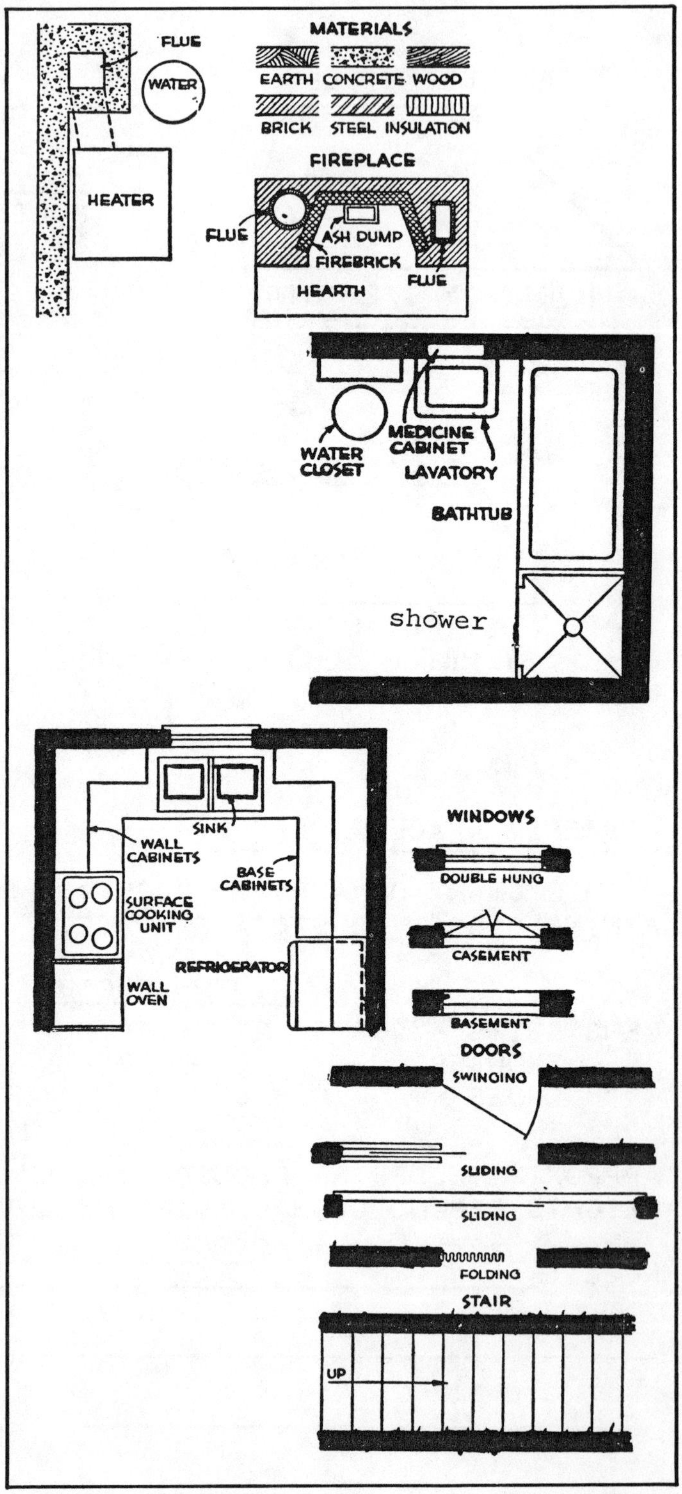

SEE: ELEVATIONS, FLOOR PLANS.

BOARD FEET MADE SIMPLE-----

THE CALCULATION OF BOARD FEET IS MADE VERY EASY BY THIS SIMPLE METHOD. A BOARD FOOT IS A PIECE OF LUMBER 1 INCH THICK, 1 FOOT WIDE AND 1 FOOT LONG. TWO WAYS TO GET THE TOTAL NUMBER OF BOARD FEET ARE AS FOLLOWS:

A. MULTIPLY THE THICKNESS IN INCHES BY THE WIDTH IN FEET BY THE LENGTH IN FEET. THUS 1" THICK X 1' WIDE X 5' LONG WOULD CONTAIN 5 BOARD FEET.

B. MULTIPLY THE SQUARE FEET OF SURFACE BY THE THICKNESS IN INCHES. THUS 1' WIDE AND 5' LONG, MULTIPLIED BY 2" THICK, WOULD BE 10 BOARD FEET.

IF YOU DOUBLE ANY SINGLE DIMENSION OF A PIECE OF LUMBER, YOU DOUBLE ITS VOLUME; DOUBLE TWO DIMENSIONS, ITS VOLUME INCREASES 4 TIMES. SEE, ALSO, *LUMBER PRICING*.

BATHROOM EQUIPMENT-----

USE THE AVERAGE DIMENSIONS OF STANDARD EQUIPMENT TO PLAN AND DETERMINE THE PROPER SIZES FOR THE BATHROOMS IN YOUR HOME.

	TYPE	WIDTH	LENGTH
BATHTUB	STANDARD	28" TO 32"	54" TO 72"
(HEIGHT VARIES)	SQUARE	42"	37"
		39"	38"
		49"	48"
LAVATORY	STANDARD	12" TO 22"	12" TO 27"
(HEIGHT: 34" TO 38")	PULLMAN	16" TO 20"	32" TO 42"
(CHILDREN 28" TO 32")	TWIN	16" TO 24"	54" TO 72"
SHOWER STALLS (HEIGHT VARIES)		36" TO 40"	36" TO 40"
WATER CLOSET STALLS	SIDE DOOR	30" TO 36"	60"
(MIN. SPACE REQUIRED)	END DOOR	30" TO 36"	57"
WATER CLOSET	SPACE REQ'D	30"	48"

STANDARD SHOWER HEAD HEIGHTS (MINIMUM):
MEN 66" WOMEN 60" FLEXIBLE TYPE 54".

BATHROOM PLANNING, FIXTURE CLEARANCES-----

THESE EASY-TO-UNDERSTAND SKETCHES WILL HELP YOU LOCATE FIXTURES AND PROVIDE MINIMUM CLEARANCE WHEN PLANNING BATHROOM LAYOUTS.

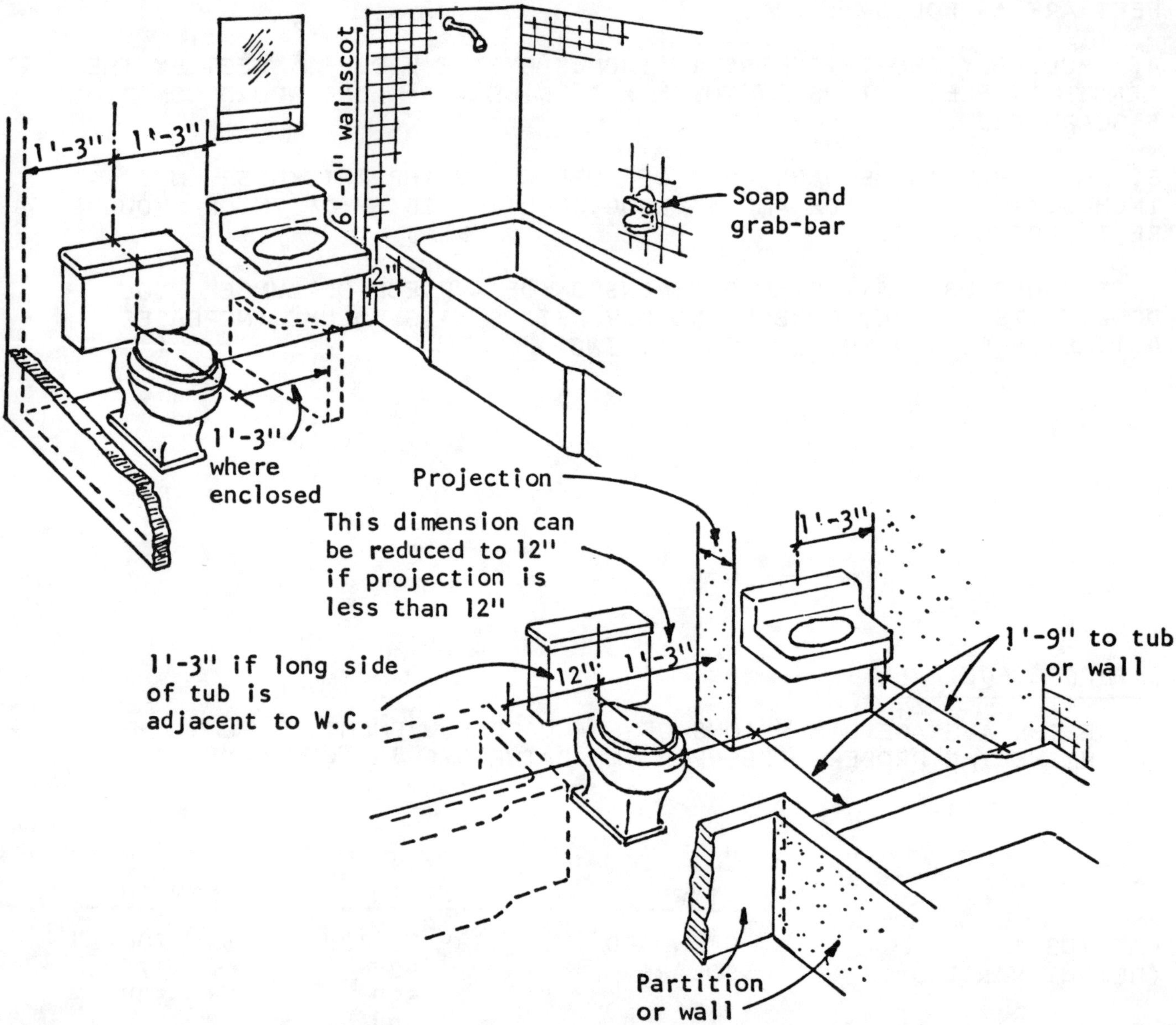

REFER TO THE BOOK: <u>SUCCESSFUL BATHROOMS</u> (STRUCTURES PUBLISHING CO.) ALSO *BATHROOM EQUIPMENT* IN THIS BOOK.

Notes: ______________________________

____

____

____

____

TOILET TYPES-----

HERE ARE THREE TYPES OF TOILETS AND THE AVERAGE DIMENSIONS. SEE PREVIOUS PAGE FOR CLEARANCES REQUIRED FOR INSTALLATION.

AVERAGE DIMENSIONS

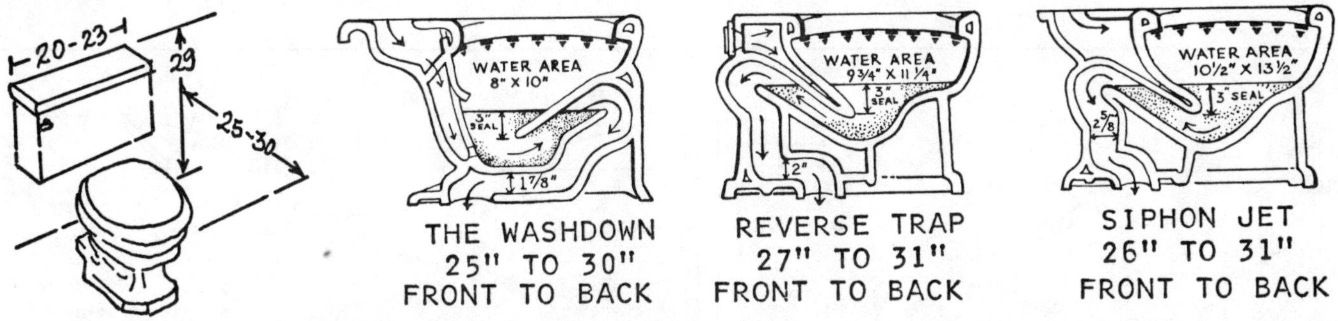

THE WASHDOWN 25" TO 30" FRONT TO BACK

REVERSE TRAP 27" TO 31" FRONT TO BACK

SIPHON JET 26" TO 31" FRONT TO BACK

COURTESY SUCCESSFUL BATHROOMS (STRUCTURES PUBLISHING CO.)

BRICK BONDS, TYPES, JOINTS-----

FACE BRICK AND BUILDING BRICK TYPES, SIZES, SHADES AND COLOR COMBINATIONS ARE IMPORTANT IN PROVIDING BEAUTIFUL AND PRACTICAL WALLS FOR HOMES. BURNED CLAY OR SHALE OR CONCRETE BRICK IN SIZES, TYPES, COLORS AND TEXTURES VARY IN DIFFERENT PARTS OF THE COUNTRY.

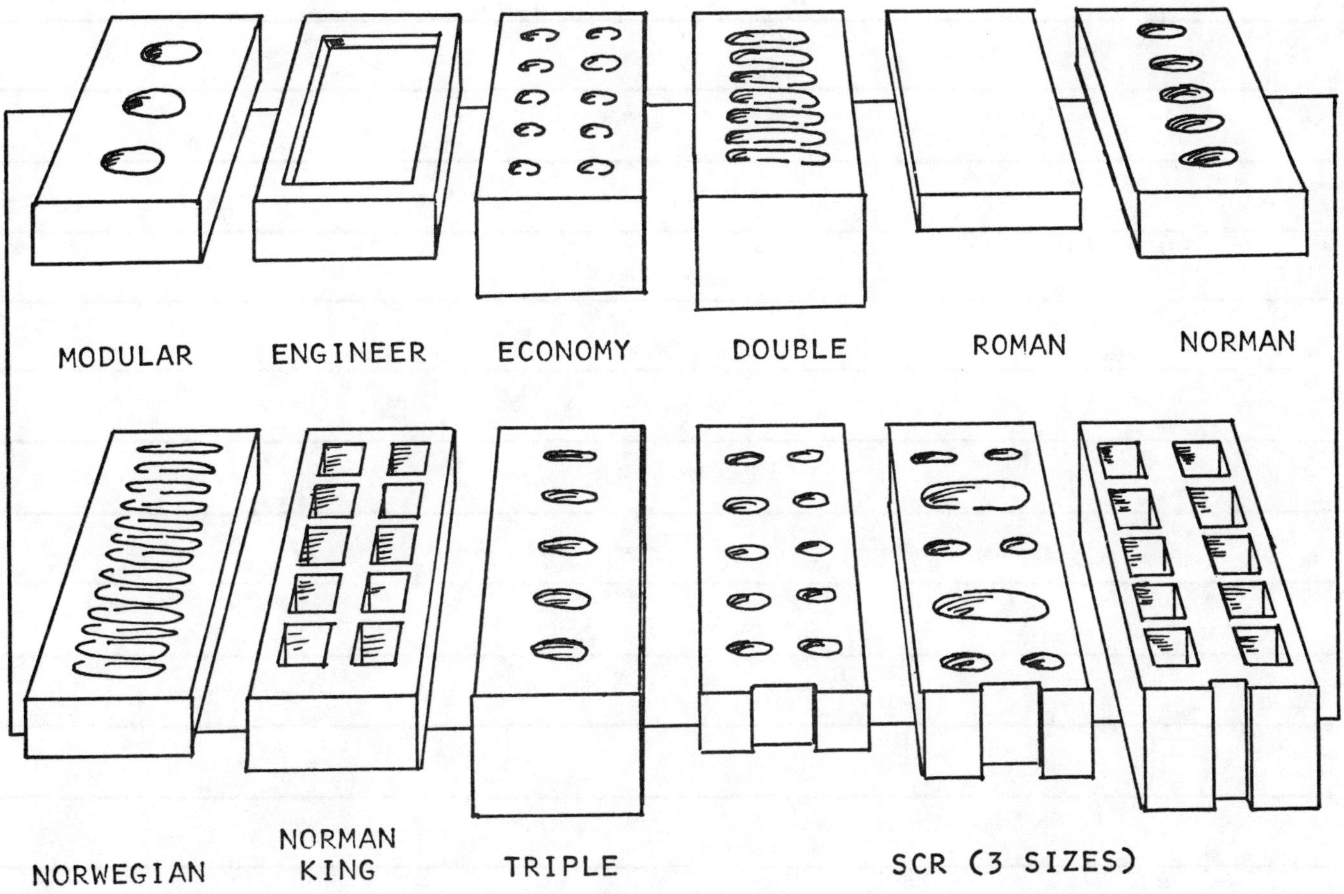

BASEMENT DOORS AND CELLAR WAYS-----

CONVENIENT OUTDOOR ENTRANCES FOR BELOW GROUND LEVEL AREAS ARE PROVIDED BY BASEMENT DOORS AND CELLAR WAYS. THEY ARE USED IN THE PLANS OF MANY NEW HOMES AND IN MODERNIZING PRESENT HOMES. TYPICAL PARTS AND AVERAGE DIMENSIONS ARE SHOWN IN THESE SKETCHES.

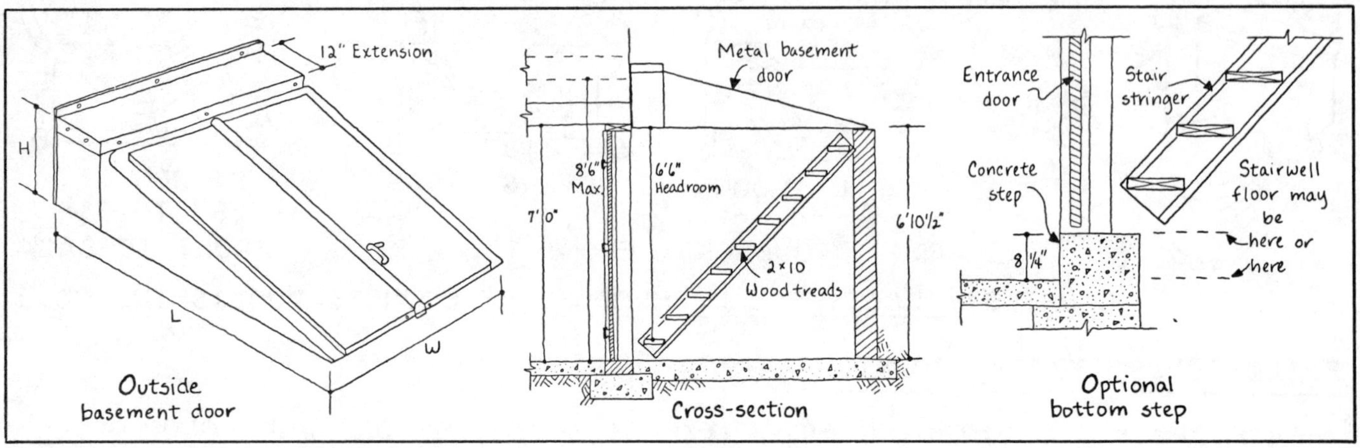

DRAWINGS COURTESY OF BILCO COMPANY.

Notes: ______________________________

BRICK BONDS-----

BRICK BONDS ARE THE DIFFERENT WAYS BRICK MAY BE LAID IN THE WALL TO PROVIDE THE DESIRED PATTERNS, WALL TEXTURES AND EFFECTS.

SOME OF THE MOST COMMONLY USED AND MOST PRACTICAL TYPES ARE SHOWN IN THE SKETCHES BELOW.

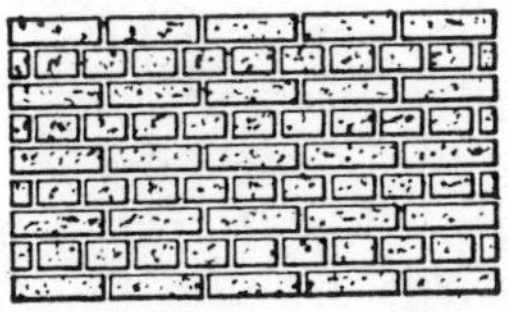
ENGLISH

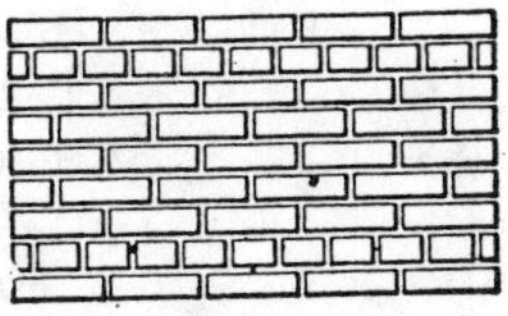
COMMON

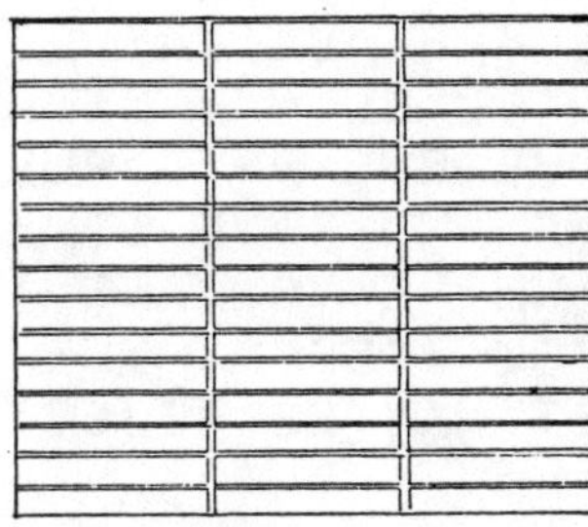
STACK BOND

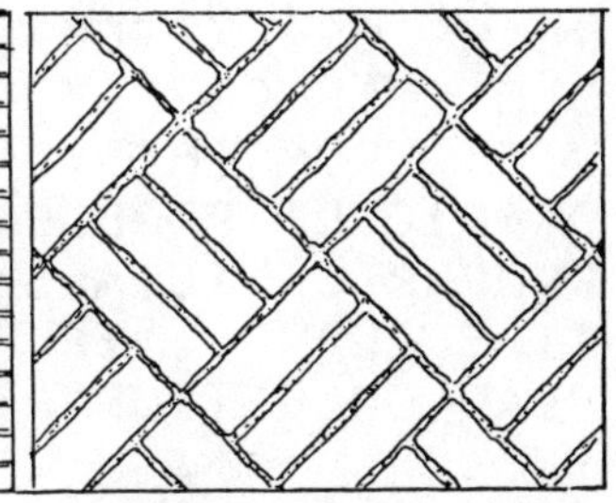
BASKET WEAVE

FLEMISH

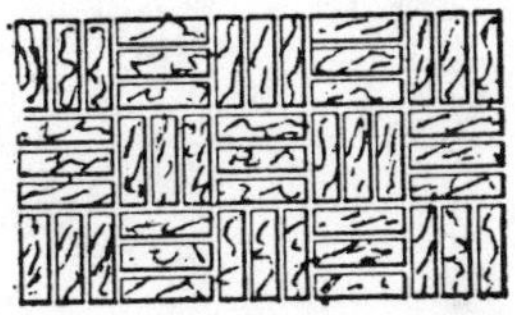
BASKET PATTERN

HERRINGBONE

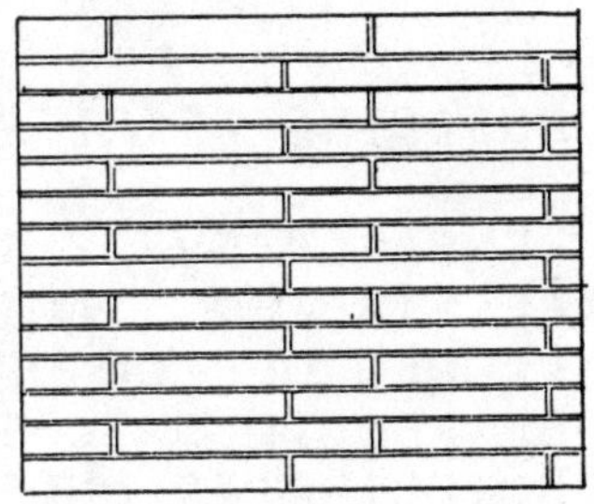
END BOND

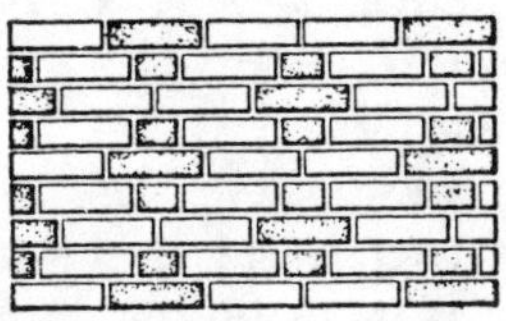
FLEMISH CROSS

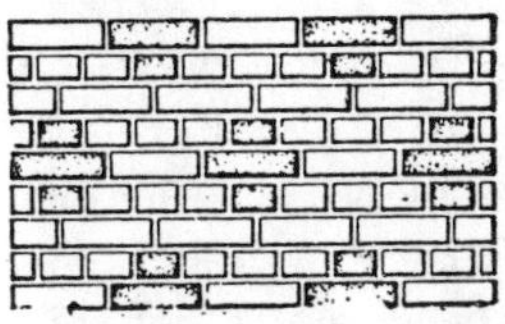
ENGLISH CROSS

BRICK AND MASONRY JOINTS-----

MASONRY JOINTS OF VARIOUS TYPES AND STYLES ARE USED TO PROVIDE THE DESIRED WALL APPEARANCE, TEXTURE AND PATTERN. MORTAR COLORS IN MANY VARIATIONS ARE ALSO AVAILABLE TO FURTHER ENHANCE THE TEXTURE AND DESIGN. THE SKETCHES HERE SHOW SOME OF THE WIDELY USED JOINTS.

SOMETIMES AFTER THE BRICK HAS BEEN SELECTED FOR A NEW HOME, THE MASON WILL PREPARE SAMPLE PANELS SHOWING THE VARIOUS TYPE JOINTS, BONDS, AND MORTARS FOR THE OWNER'S APPROVAL.

STRUCK

RAKED

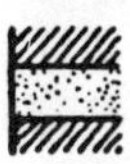
FLUSH

VEE

CONCAVE

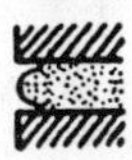
BEADED

WEATHERED

BUILDING CODES-----

USUALLY RESTRICTIONS ARE IMPOSED ON ALL KINDS AND TYPES OF BUILDING AND REMODELING. CONTROL IS MAINTAINED BY A DIVISION OF THE LOCAL OR STATE GOVERNMENT. CITIES, TOWNS AND VILLAGES, COUNTIES, BOROUGHS, OR TOWNSHIPS HAVE OFFICES OR AUTHORITIES WHO HANDLE THIS MATTER. IF YOU PLAN TO BUILD OR DO EXTENSIVE REMODELING OR RESTORATION, CONTACT THIS OFFICE AS SOON AS YOUR PLANS OR SKETCHES ARE AVAILABLE. IF YOU ARE PREPARING YOUR OWN SKETCHES OR LAYOUT, YOU MAY BE ABLE TO GET EXPERT HELP AND ADVICE FROM THIS SOURCE. AS A RULE, IT IS MUCH EASIER TO HANDLE DIFFICULT PROBLEMS AND SAVE TIME AND MONEY IF YOU WORK WITH THE PROPER OFFICIALS WHILE YOU ARE IN THE PROCESS OF MAKING TENTATIVE PLANS. IF YOU PRESENT THE PROBLEM OR PICTURE AS A COMPLETED PLAN OR SET-UP, IT MAY BE REJECTED ENTIRELY.

BUILDING PERMITS-----

NEARLY EVERY TYPE PROJECT NOW REQUIRES A PERMIT BEFORE WORK CAN BE STARTED. AUTHORITY MUST BE ACQUIRED AND APPROVED BY THE PROPER OFFICE OR OFFICIALS. WHEN YOU APPLY FOR A PERMIT YOU WILL BE REQUIRED TO SUBMIT PLANS AND SPECIFICATIONS IF A NEW HOME OR BUILDING IS UNDER CONSIDERATION. USUALLY COMPLETE FOUNDATION AND FLOOR PLANS WITH TWO ELEVATIONS, SECTION AND PLOT PLAN WILL BE REQUIRED. ON EXTENSIVE ALTERATIONS TO THE EXTERIOR OR MODERNIZATION OF A PROJECT, PLANS OR COMPLETE SCHEDULE OF THE WORK WILL ALSO REQUIRE A PERMIT. THE COST OF A PERMIT IS NOMINAL AND USUALLY BASED ON COST, WITH THE RATE VARYING IN DIFFERENT PARTS OF THE COUNTRY. THE PERMIT PROVIDES FOR THE PERIODIC INSPECTION OF THE WORK AS IT PROGRESSES AND IS ASSURANCE THAT THE WORK WILL MEET THE VARIOUS CODES. IN MOST CASES THE INSURANCE COMPANIES REQUIRE CODE COMPLIANCE INSOFAR AS WIRING, PLUMBING AND HEATING ARE CONCERNED. GENERALLY, IF A CONTRACTOR OR BUILDER IS HANDLING THE WORK, HE WILL HANDLE THE MATTER OF PERMITS FOR YOU.

BUILDING RESTRICTIONS-----

RESTRICTIONS ON BUILDING ARE COMMONLY FOUND IN THE DEED FOR THE LAND. BASICALLY THEY LIMIT THE USE OF THE LAND. OFTEN THEY ARE SHOWN IN THE SALES CONTRACT OR AGREEMENT. USUALLY THE RESTRICTIONS ARE PLACED BY THE DEVELOPERS OF THE LAND UNLESS THEY CARRY THROUGH FROM VERY EARLY OWNERS. THEY MAY ALSO BE INSTITUTED BY CITIES, TOWNS, COUNTIES OR TOWNSHIPS THROUGH ORDINANCES OR BUILDING CODES. THEY MAY COVER BUILDING LOCATIONS, SETBACKS, SIZES, TYPES AND MANY OTHER UNUSUAL CONDITIONS. THEY MAY ALSO TAKE THE FORM OF CONDITIONS OR COVENANTS. YOU CAN CHECK THE RESTRICTIONS ON ANY PROPERTY OR HAVE IT DONE BY A TITLE SEARCH.

BUYER-SELLER EXPENSES-----

WHEN REAL PROPERTY IS BOUGHT OR SOLD, THE VARIOUS COSTS INCURRED BY BOTH THE BUYER AND SELLER MAY VARY IN DIFFERENT LOCALITIES, DEPENDENT UPON THE CONDITIONS OF THE TRANSACTION, LENDING AGENCY, REALTOR OR AGREEMENTS BY BOTH PARTIES. YOU SHOULD ALWAYS HAVE A FULL AND COMPLETE UNDERSTANDING REGARDING TITLE INSURANCE, SURVEYS, TAXES, CLOSING COSTS, LOAN FEES, UTILITIES, PROPERTY INSURANCE, RECORDING, REVENUE STAMPS AND SALES TAX.

BUYING A HOME-----

IF YOU ARE BUYING A USED HOME, CAREFULLY CHECK THE ROOM SIZES AND THE ADAPTABILITY OF YOUR PRESENT FURNITURE OR REPLACEMENTS. THE EASIEST AND MOST SIMPLE WAY TO DO THIS IS TO MAKE SCALED SKETCHES OF THE ROOMS - LOCATING DOORS, WINDOWS, ETC., AND THEN FIT SCALED SKETCHES OF YOUR FURNITURE INTO THE AREAS. THIS MAY BE DONE QUICKLY BY THE USE OF THE SKETCH SHEETS ELSEWHERE IN THIS BOOK. TEMPLATES FOR DRAWING FURNITURE, BATHS, KITCHENS AND REMODELING ARE AVAILABLE AT MOST OFFICE SUPPLY STORES. THEY SIMPLIFY THE SKETCHING PROBLEM AND ARE EASILY USED BY ANYONE. IT IS ALSO ADVISABLE TO CHECK THE SIZES OF AVAILABLE CLOSETS, CUPBOARDS AND STORAGE SPACE. MAKE SURE THEY WILL MEET YOUR REQUIREMENTS OR ARE ALTERED TO SUIT YOUR NEEDS. IF REMODELING IS INDICATED, SKETCHES OF THE PROPOSED CHANGES MAY BE MADE IN THE MANNER INDICATED ABOVE. FOR MODERNIZING KITCHENS AND BATHS, THE 1/2" SCALE TEMPLATE IS SUGGESTED TO PROVIDE A LARGE AND EASY TO READ SKETCH.

STICKING DOORS AND WINDOWS, BROKEN LOCKS, LEAKY FAUCETS AND TOILET, BROKEN SWITCHES, MISSING SWITCH PLATES, LIGHT BULBS, POOR PAINT AND FINISH, DIRT AND UNTIDY CONDITIONS ALL INDICATE NEGLIGENCE IN MAINTENANCE AND DETRACT FROM THE VALUE OF A USED HOME. THESE ITEMS SHOULD BE CORRECTED BEFORE BUYING UNLESS AN ALLOWANCE IS MADE. IN MANY CASES, A GOOD, SOUND HOUSE THAT REQUIRES EXTENSIVE WORK AND IMPROVEMENT MAY BE PURCHASED AT AN ATTRACTIVE PRICE WITH THE IDEA OF UPDATING AND MODERNIZING. WITH THIS IN VIEW, ALTERATION COSTS SHOULD BE CAREFULLY CHECKED BEFORE PURCHASING TO INSURE A REASONABLE TOTAL COST AND ULTIMATE VALUE. CODES AND LOCAL BUILDING REGULATIONS SHOULD BE CHECKED BEFORE EXTENSIVE IMPROVEMENTS ARE MADE.

IN A USED HOME, ALWAYS CHECK THE PLUMBING, HEATING, WIRING, ROOF, BATH, KITCHEN AND BASEMENT AS WELL AS THE WATER HEATER AND FUSE BOX FOR FAULTY AND DEFECTIVE EQUIPMENT. OFTEN A NON-INTERESTED PARTY EXPERIENCED IN THE BUSINESS IS RETAINED TO CHECK ON THESE AND OTHER ITEMS. USUALLY THE OWNER OR AGENT WILL PROVIDE COPIES OF UTILITY AND TAX BILLS SO YOU MAY EVALUATE THESE EXPENSES. FUTURE ASSESSMENTS OR CHANGES IN TAXING METHODS SHOULD BE CHECKED IN CASE EXTENSIVE TAX CHANGES ARE PROPOSED. A VERY IMPORTANT ITEM IS THE MATTER OF A TERMITE CHECK IN AREAS WHERE TERMITES ARE PREVALENT. QUITE OFTEN THE TERMITE CHECK IS PART OF THE SALES CONTRACT. ANOTHER MATTER OF IMPORTANCE IS THE LOCKS.

BUYING A HOME (continued)-----

IT IS ADVISABLE TO HAVE ALL THE DOOR LOCKS FITTED WITH NEW CYLINDERS AND NEW KEYS FOR SECURITY REASONS. A CLEAR TITLE IS ALWAYS CONSIDERED NECESSARY AND USUALLY IS HANDLED BY A TITLE COMPANY RETAINED BY THE SELLER OR THE LENDING AGENCY IF A MORTGAGE IS REQUIRED. THIS INSURES THE BUYER AGAINST LIENS, MORTGAGES OR OTHER CLAIMS THAT MIGHT BE HELD AGAINST THE PROPERTY. MAKE SURE THE BUILDING IS FULLY INSURED AGAINST DAMAGE JUST AS SOON AS THE TITLE IS IN YOUR NAME. THE CONTENTS SHOULD ALSO BE FULLY INSURED, EFFECTIVE THE DAY YOU MOVE YOUR FURNITURE INTO THE HOME.

Notes: ______________________________

BUYING OR SELLING REAL ESTATE-----

IF YOU ARE BUYING OR SELLING PROPERTY YOU WILL ENCOUNTER MANY TERMS THAT CAN BE CONFUSING. THIS LIST OF COMMONLY USED TERMS AND THEIR DEFINITIONS WILL ENABLE YOU TO DISCUSS AND UNDERSTAND THE TERMINOLOGY USED BY REALTORS, LAWYERS, APPRAISERS AND LENDING ORGANIZATIONS.

ABSTRACT. THE HISTORY OF A PARCEL SHOWING ALL THE TRANSACTIONS OF ITS CONTINUED OWNERSHIP.

APPRAISAL. THE VALUE OF THE PROPERTY AS DETERMINED BY THE OPINION OF EXPERTS, SOMETIMES DETERMINED BY RECENT SALES OF SIMILAR TYPE PROPERTY, BY TAXING VALUES, OR LENDING AGENCIES.

CONTINGENCY. A CONTINGENCY COVERS PROPERTY WHICH IS SOLD OR BOUGHT WITH THE UNDERSTANDING THAT THE DEAL WILL BE CONSUMMATED IF AND WHEN ANOTHER SALE, PURCHASE OR CONDITION IS MET WITHIN A CERTAIN STIPULATED DATE.

CONVENTIONAL MORTGAGE LOAN. A LOAN HANDLED BY LENDING INSTITUTIONS WITHOUT APPROVAL OR GUARANTEE OF EITHER FHA (FEDERAL HOUSING ADMINISTRATION) OR VA (VETERANS ADMINISTRATION).

DEED. AN OFFICIAL PAPER PROPERLY EXECUTED TO SHOW THE TRANSFER OF REAL PROPERTY FROM ONE OWNER TO ANOTHER.

EQUITY. THE ACTUAL AMOUNT OF MONEY THE OWNER HAS IN A PROPERTY. ALSO THE DIFFERENCE BETWEEN THE SELLING PRICE, AFTER SALES EXPENSES, AND THE AMOUNT THE OWNER OWES ON THE PROPERTY.

FHA LOAN. A LOAN APPROVED BY THE FEDERAL HOUSING ADMINISTRATION AND INSURING THE LENDING INSTITUTION AGAINST LOSS DUE TO THE FAILURE OF THE BUYER TO FULFILL HIS OBLIGATIONS.

FMHA LOAN. A DIRECT LOAN MADE BY THE FARMERS HOME ADMINISTRATION.

LISTING. PROPERTY ASSIGNED TO A REAL ESTATE AGENT OR BROKER FOR SALE AS APPROVED BY THE OWNER. TERMS AND CONDITIONS ARE GENERALLY HANDLED BY THE USE OF AN APPROVED AND SIGNED AGREEMENT OR LISTING FORM.

LOAN COMMITMENT. THE APPROVAL OF A LENDING AGENCY TO LOAN A GIVEN OR STATED AMOUNT ON A SPECIFIC PROPERTY. IT USUALLY INDICATES THE INTEREST RATE, LENGTH OF LOAN AND MONTHLY PAYMENTS.

OFFER TO BUY. AN OFFER AGREEMENT MADE BY A POTENTIAL BUYER STATING PRICE, TERMS, SALE CONDITIONS, OCCUPANCY, ETC.

REALTOR. LICENSED AGENTS OR BROKERS AND MEMBERS OF THE BOARD OF REALTORS.

REAL ESTATE BROKER OR AGENT. PERSONS OFFICIALLY LICENSED TO DEAL IN REAL PROPERTY.

REAL ESTATE CONTRACT. AS COMPARED TO AN OFFER TO BUY, A CONTRACT SETS FORTH ALL THE SPECIFIC TERMS, CONDITIONS OF PAYMENT, AND OTHER PERTINENT ITEMS AS AGREED UPON BY BOTH PARTIES.

TITLE INSURANCE. PROVIDES FOR CLEAR OWNERSHIP OF THE PROPERTY COVERED. IT BINDS THE INSURANCE COMPANY TO PAY THE INSURED FOR LOSSES INCURRED DUE TO DEFECTS IN THE TITLE AND GUARANTEES A CLEAR TITLE FOR THE PROPERTY.

VA LOAN. A LOAN APPROVED BY THE VETERANS ADMINISTRATION AND MADE TO VETERANS.

SELECTING A QUALIFIED BUILDER-----

AFTER YOU HAVE SELECTED YOUR NEW HOMESITE AND SECURED YOUR PLANS, YOU WILL BE READY TO SELECT A QUALIFIED BUILDER TO DO THE WORK, UNLESS YOU ARE A DO-IT-YOURSELFER—OR EXPECT TO ACT AS YOUR OWN CONTRACTOR.

YOU MAY DEVELOP A LIST OF POTENTIAL BUILDERS IN YOUR AREA BY TALKING TO FRIENDS, CHECKING NEW HOMES OF THE TYPE YOU DESIRE, LOOKING AT HOMES UNDER CONSTRUCTION, ADVERTISEMENTS, TELEPHONE DIRECTORIES, CONSULTING WITH LENDING AGENCIES AND OTHERS. IN MOST AREAS YOU WILL FIND A GROUP KNOWN AS THE HOME BUILDERS ASSOCIATION AND/OR THE CONTRACTORS ASSOCIATION. EITHER OF THESE WILL PROVIDE NAMES OF REPUTABLE BUILDERS.

AFTER YOU SELECT BIDDERS AND SECURE A SATISFACTORY CONTRACT PRICE YOU WILL THEN BE READY TO DO FURTHER CHECKING ON THE BUILDER YOU CHOOSE. THIS MAY BE DONE BY CHECKING THE FIRM'S RELIABILITY THROUGH THE BANKS, LENDING AGENCIES OR THE BETTER BUSINESS BUREAU. SOME LOCALITIES REQUIRE THAT CONTRACTORS BE LICENSED---IN WHICH CASE YOU CAN CHECK WITH THE BUILDING DEPARTMENTS. SOME OWNERS REQUIRE A BOND OR GUARANTEE INSURING THE JOB AGAINST FAILURE ON THE PART OF THE CONTRACTOR, LIENS OR OTHER PROBLEMS.

THE PLUMBING, HEATING AND AIR CONDITIONING AND ELECTRICAL PORTIONS OF MOST CONTRACTS ARE HANDLED BY SUB-CONTRACTORS. AS THE OWNER YOU SHOULD ALSO CHECK THESE SUBS AS WELL AS OTHER CRAFTS THAT MAY BE SUBLET.

BE SURE TO HAVE A GOOD, COMPLETE UNDERSTANDING, BY AGREEMENT OR CONTRACT IN WRITING, COVERING ALL CONDITIONS SUCH AS EXTRAS, ALTERNATES, SPECIFICATIONS, INSURANCE AND PAYMENTS BEFORE YOU SIGN YOUR CONTRACT.

PROPER ATTENTION TO THESE IMPORTANT MATTERS BEFORE YOUR WORK IS STARTED WILL PAY OFF IN A PROPER RELATIONSHIP WITH YOUR BUILDER AND HIS SUBS AND CRAFTSMEN AND WILL RESULT IN A BETTER HOME FOR YOU.

BROOM CLOSETS-----

BROOM OR UTILITY CLOSETS FOR THE STORAGE OF HOUSEHOLD CLEANING EQUIPMENT ARE USUALLY LOCATED IN KITCHEN, LAUNDRY OR UTILITY ROOMS. TYPICAL DIMENSIONS ARE AS FOLLOWS:

WIDTH: 15" TO 48"
DEPTH: 23" TO 25"
HEIGHT: 72" TO 84" (OR CEILING HEIGHT WITH SHELVES AT THE TOP).

CLOSETS-----

CLOSETS AND CABINETS OF VARIOUS TYPES AND SIZES ARE VITAL IN EVERY HOME. THEY UTILIZE SPACE TO THE BEST ADVANTAGE AND PROVIDE STORAGE FOR MANY UNUSUAL AND EXTRA ITEMS THAT MUST BE KEPT IN THE HOME.

WHERE CLOTHES RODS AND SHELVES ARE TO BE USED, THE RODS AND SHELVES SHOULD BE MOUNTED AT LEAST FIVE FEET CLEAR OF THE FLOOR OR OTHER OBSTRUCTIONS BUT NOT MORE THAN SIX AND ONE HALF FEET ABOVE THE FLOOR LEVEL.

CLOSETS IN CHILDRENS' ROOMS SHOULD HAVE ADJUSTABLE HEIGHTS—FOR PRACTICAL USE AS THE CHILDREN GROW OR FOR OTHER USES IF REQUIRED.

APPROXIMATE DEPTHS REQUIRED: 24" TO 30".
HEIGHTS DEPEND UPON CEILING HEIGHTS, USUALLY 90" TO 96".

Notes:

CARPET TERMS-----

THE FOLLOWING ARE TERMS, DESCRIPTIONS AND EXPLANATIONS COVERING MOST TYPES OF CARPETING:

BACKING - FOUNDATION ON UNDERSIDE OF CARPET INTO WHICH PILE IS SECURED. USUALLY MADE OF JUTE, COTTON OR KRAFT CORD.

BLEND - A COMBINATION OF TWO OR MORE FIBERS. EACH FIBER IS LISTED IN ORDER OF PREDOMINANCE.

BROADLOOM - A WIDE WIDTH, SEAMLESS CARPET OF ANY WEAVE MADE ON LOOMS 6 TO 18 FEET WIDE.

CRUSH RECOVERY - THE ABILITY OF CARPET PILE TO REGAIN ITS ORIGINAL HEIGHT AND TEXTURE AFTER CRUSHING BY FURNITURE AND CONSTANT TRAFFIC.

FIBERS - THE INDIVIDUAL STRANDS WHICH ARE SPUN TOGETHER TO MAKE THE CARPET PILE YARN.

FLUFFING - A FUZZ THAT CLINGS TO THE PILE DURING SHEARING AND GRADUALLY SHEDS.

HEAT SET - THE YARN TWIST THAT HAS BEEN MADE PERMANENT THROUGH APPLICATION OF HIGH HEAT AND MOISTURE.

PILE - YARN TUFTS THAT STAND ERECT FROM THE CARPET BACKING AND FORM THE SURFACE WITH LOOP OR CUT ENDS. ALSO CALLED "FACE YARN."

PILLING - TINY BALLS OR "PILLS" OF FIBER WHICH ROLL UP ON SURFACE OF CARPET AND REMAIN ATTACHED TO THE PILE.

TEXTURE - SURFACE INTEREST ACHIEVED IN VARIOUS WAYS, SUCH AS VARIATION OF PILE, HEIGHTS, COMBINATION OF LOOP AND CUT PILE, TWIST AND PLAIN WEAVE, OR CARVED DESIGNS.

TUFTED - TUFTED CARPETS ARE MADE BY NEEDLING PILE YARNS INTO A PREVIOUSLY WOVEN BACKING OF JUTE OR COTTON.

TWIST - THE DIRECTION AND SHAPE THE YARN HAS BEEN GIVEN TO PRODUCE A PARTICULAR TEXTURE EFFECT.

WAFT OR "FILLING" - THE YARN WHICH RUNS ACROSS THE WIDTH OF A WOVEN CARPET.

WARP - THE YARN WHICH RUNS THE LENGTH OF THE CARPET. PILE YARN IN WOVEN TYPES.

WIDTHS - STANDARD WIDTHS ARE 27 INCHES, 9 FEET, 12 FEET, 15 FEET, 18 FEET, ALSO SOME 3 FEET AND 6 FEET. ANY WIDTH IS AVAILABLE IN A CUSTOM MADE CARPET.

WILTON, VELVET, AXMINSTER - THREE PRINCIPAL WEAVES USED TO OBTAIN DIFFERENT STYLES IN WOVEN CARPETS.

CARPET TYPES-----

WALL TO WALL CARPETING IS USED EXTENSIVELY IN MANY PARTS OF HOMES TODAY. IT IS USUALLY SOLD BY THE SQUARE YARD. NORMALLY, IT IS MADE IN 6, 12 AND 15 FOOT WIDTHS. WHEN SELECTING CARPET IT IS IMPORTANT TO MAKE SURE IT HAS A STRONG BACKING. GOOD THICK PADDING LENGTHENS THE LIFE AND PROVIDES UNDERFOOT COMFORT. REGULATIONS REQUIRE THAT CARPET LABELS SHOW THE WEIGHT PERCENTAGE OF EACH FIBER USED IN BLEND COMBINATIONS.

THE COMMON NAMES AND A BRIEF DESCRIPTION OF BASIC CARPET FIBERS ARE SHOWN BELOW.

ACRYLIC - GOOD WEARING, RESILIENT AND COLOR FAST. SOFT APPEARANCE.

COTTON - POOR RESILIENCE, WASHABLE, FAIR SOIL RESISTANCE, GOOD WEARING.

MODACRYLIC - MODERATE RESILIENCE AND GOOD WEAR RESISTANCE.

NYLON - GOOD COLORFASTNESS AND RESILIENCE. EXCELLENT WEARING QUALITIES.

POLYESTER - LUXURIOUS APPEARANCE AND GOOD WEAR. SOFT, WITH COLORFASTNESS.

POLYPROPYLENE - RESISTANT TO MOISTURE. INDOOR-OUTDOOR TYPE, COLORFAST. ALSO KNOWN AS "OLEFIN."

RAYON - WASHABLE, USED AS SCATTER TYPES. FAIR RESISTANCE TO WEAR AND STAINS.

WOOL - SOFT, WARM APPEARANCE, EXCELLENT RESILIENCE, GOOD RESISTANCE TO SOIL AND WEAR.

CARPET PILE TYPES AND STYLES ARE SHOWN IN THE SKETCHES BELOW.

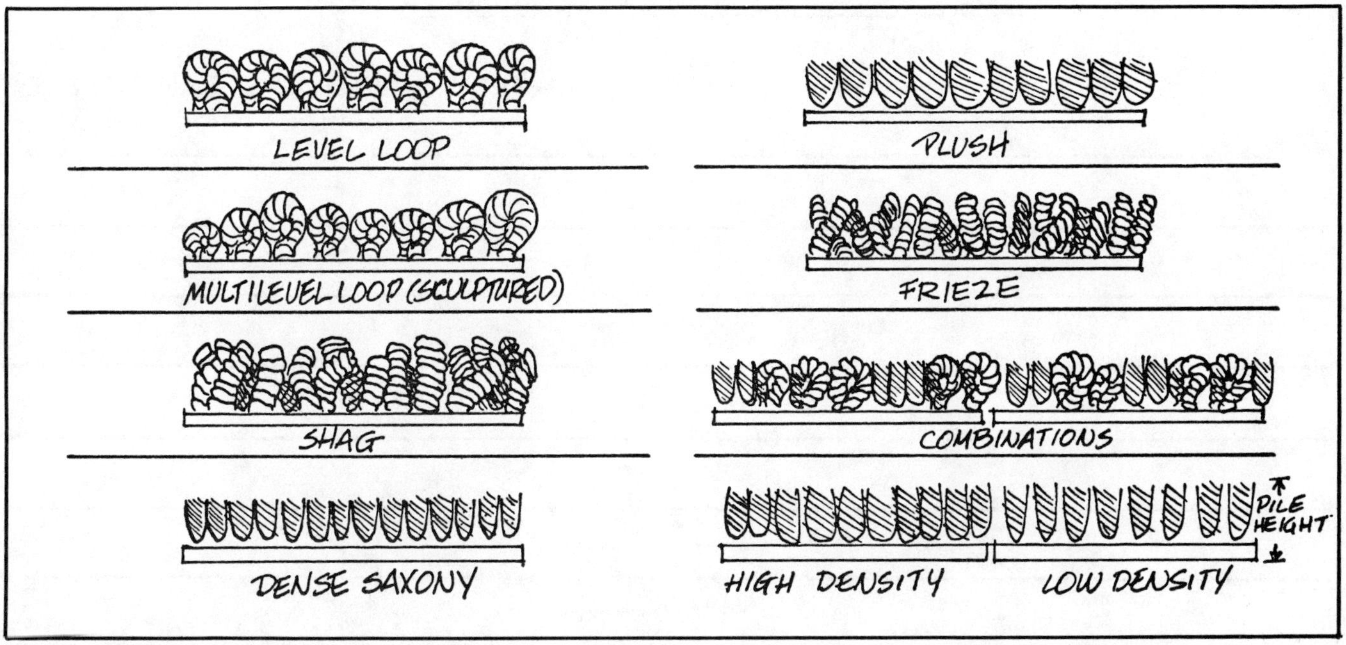

CERAMIC TILE-----

CERAMIC TILE PLAYS AN IMPORTANT PART IN MANY HOMES TODAY. IT IS AVAILABLE IN MANY SIZES, TYPES AND SHADES TO SUIT MOST DECOR CONDITIONS. DEPENDING ON THE TYPE, IT MAY BE USED ON WALLS, FLOORS, SINKTOPS, ETC.

TYPE	SIZES
DOTS	1/2" - 11/32"
SQUARES	3/4" - 1-1/16" - 1-9/16" - 2-3/16" - 4-1/4" - 6"
OBLONG	3/4" X 1-9/16" 1/2" X 1-1/16" 1" X 2" 1-1/16" X 2-3/16" 4-1/4" X 6" 6" X 9"
HEXAGONAL	1" - 1-1/4" - 2" - 3"
OCTAGONAL	2-3/16"
PENTAGON	2-3/16"
THICKNESS	1/4" - 5/16" - 3/8" - 1/2"

VERY SMALL TILE IN REGULAR AND UNIQUE PATTERNS IS AVAILABLE IN SELF-SPACED, GLUED ON SLABS, APPROXIMATELY 1' SQUARE.

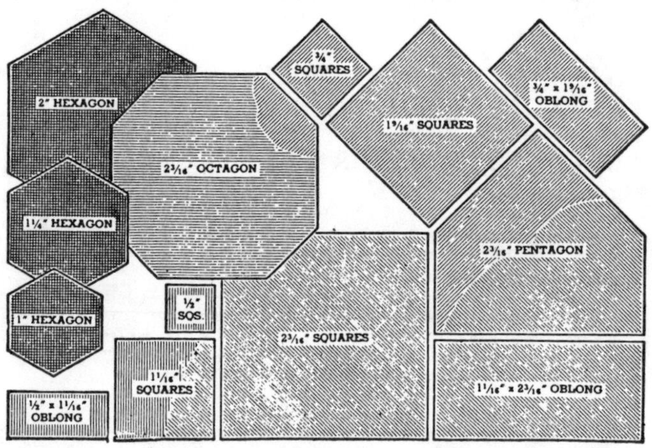

COURTESY AMERICAN OLEAN TILE CO.

Notes:

COLORS-----

THE USE OF THE VARIOUS AND PROPER COLORS IN THE SCHEME OF DECORATING WILL HELP CONSERVE ENERGY.

THE REFLECTION OF LIGHT OFF DIFFERENT COLORS IS SHOWN IN THE CHART BELOW.

THE USE OF THE PROPER COLORS WILL PROVE HELPFUL IN MANY WAYS AND WILL ALSO PROVIDE MORE COMFORTABLE LIVING CONDITIONS.

BLACK	2%	GRAY (MEDIUM)	50%
BLUE (MEDIUM)	35%	GREEN (SEA)	67%
BROWN (LIGHT)	20%	GREEN (APPLE)	30%
BUFF (LIGHT)	74%	IVORY	82%
CORAL	64%	ORCHID (LIGHT)	79%
CREAM (LIGHT)	77%	YELLOW (MEDIUM)	75%
CREAM (MEDIUM)	70%	WHITE (GLOSS)	90%
GRAY (LIGHT)	52%	WHITE (DULL)	80%

CHIMNEY HEIGHTS-----

CARE SHOULD ALWAYS BE EXERCISED IN THE LOCATION AND HEIGHTS OF CHIMNEYS. ASSUMING THE SIZE HAS BEEN CORRECTLY CALCULATED, PROPER DRAFT WILL BE DETERMINED BY THE HEIGHT ABOVE THE ROOF OR OTHER OBSTRUCTIONS NEARBY. TREES AND HIGHER ROOFS CAN CAUSE DOWNDRAFTS AS SHOWN HERE.

PREVAILING WINDS CAN ALSO AFFECT THE PERFORMANCE OF A CHIMNEY AND CAUSE SMOKING OR POOR DRAFT CONDITIONS. CHECK ALL THESE CONDITIONS BEFORE LOCATING, BUILDING OR REMODELING CHIMNEYS.

COURTESY BOOK OF SUCCESSFUL FIREPLACES (STRUCTURES PUBLISHING CO.)

SEE *CHIMNEYS, FIREPLACES* IN THIS BOOK.
ALSO, BOOK OF SUCCESSFUL FIREPLACES (STRUCTURES PUBLISHING COMPANY)

CHIMNEYS-----

THERE ARE MANY PARTS OF A CHIMNEY, AS SHOWN BY THE DRAWING OF THE SECTION OF TYPICAL CHIMNEY CONSTRUCTION ON THE NEXT PAGE. IT SHOWS THE VARIOUS COMPONENT PARTS AND THEIR LOCATION.

(1) CHIMNEY CAP - AVAILABLE PRECAST FOR SOME STANDARD FLUE SIZES OR CAST IN PLACE. NOTE THAT THE LINER PROJECTS THROUGH THE CAP SEVERAL INCHES.

(2) FLUE AND FLUE LINING - THE AREA OF THE FLUE SHOULD EQUAL ONE-TWELFTH TO ONE-EIGHTH THE AREA OF THE OPENING OF THE FIRE-PLACE (WIDTH TIMES HEIGHT). LINING CONSISTS OF PROPERLY SIZED CLAY FLUE TILE SECTIONS.

(3) SMOKE CHAMBER - TOGETHER WITH THE SMOKE SHELF, THIS AREA IS IMPORTANT TO A SMOKE FREE FIRE. BOTH SIDES SLOPE TO THE FLUE AND IT IS IMPORTANT THAT THEY SLOPE IDENTICALLY, OTHERWISE THE FIRE WILL BURN ON ONE SIDE OF THE FIREBOX ONLY.

THE ENTIRE SMOKE CHAMBER AND SMOKE SHELF IS PARGETED WITH FIRE CLAY MORTAR (REFRACTORY MORTAR) OR TYPE "S" MORTAR ONE-HALF INCH THICK.

(4) SMOKE SHELF - THIS IS A HORIZONTAL SHELF, USUALLY CONCAVE AND EXTENDING BACKWARD FROM THE REAR OF THE THROAT OR DAMPER TO THE REAR FLUE WALL. IT DIRECTS COLD AIR DOWNDRAFTS WHICH ARE PRESENT IN THE EARLY STAGES OF THE FIRE, CAUSING THEM TO EDDY AND DRIFT UPWARD WITH THE RISING AIR CURRENTS.

(5) THROAT AND DAMPER - THESE PARTS ARE USUALLY ONE AND THE SAME. THE DAMPER IS CAPABLE OF BEING OPENED AND CLOSED GRADUALLY TO CONTROL THE DRAFT AND KEEP OUT COLD AIR WHEN THE FIREPLACE IS NOT IN USE. THE OPENING IN SQUARE INCHES SHOULD BE AT LEAST 90% OF THAT REQUIRED FOR THE FLUE.

(6) FIREBOX - THIS IS THE AREA THAT COMES ALIVE WITH DANCING FLAMES AND GLEAMING EMBERS. TO DO THIS THE FIREBOX MUST BE CORRECTLY PROPORTIONED, SEALED, VENTED AND WELL CONSTRUCTED.

(7) PREBUILT METAL CIRCULATING FIREPLACE UNITS OF VARIOUS TYPES AND SIZES ARE OFTEN BUILT INTO THE CHIMNEY.

(8) HEARTH - THE INNER HEARTH IS THAT PORTION WITHIN THE FIRE AREA AND IS USUALLY BUILT OF FIRE BRICK BUT MAY BE OTHER TYPES OF HARD BRICK, CONCRETE, STONE, TILE OR OTHER NON-COMBUSTIBLE HEAT-RESISTANT MATERIALS. THE OUTER HEARTH IS BUILT OF THE SAME TYPE MATERIALS AND SHOULD EXTEND A MINIMUM OF EIGHT INCHES ON EACH SIDE OF THE FIREPLACE OPENING AND SIXTEEN INCHES IN FRONT. (*Note: These figures are twelve inches and eighteen inches in areas covered by the Uniform Building Code.*)

(9) ASHPIT - THIS IS THE HOLLOW SPACE BELOW THE HEARTH INTO WHICH ASHES FALL THROUGH THE ASH DUMP DOOR LOCATED IN THE HEARTH. A METAL DOOR IS PROVIDED IN THE ASHPIT FOR THE OCCASIONAL REMOVAL OF ASHES. IN BASEMENTLESS HOMES, PARTICULARLY THOSE BUILT ON A CONCRETE SLAB, IT MAY NOT BE FEASIBLE TO PROVIDE

AN ASHPIT UNLESS THE HEARTH HAS BEEN RAISED. IN THIS CASE THE ASHPIT DOOR FACES OUTSIDE.

(10) FOUNDATION - CONSULT YOUR LOCAL BUILDING CODE, SINCE THESE CODES DIFFER ACCORDING TO EXISTING SOIL AND MOISTURE CONDITIONS IN INDIVIDUAL AREAS. IF TOTAL WEIGHT IS NEEDED TO COMPUTE THE DEPTH AND RISE OF THE FOUNDATION REQUIRED, FIGURE BRICK AT 130 LBS. AND CONCRETE AT 150 LBS. PER CUBIC FOOT. FOR CUBIC FOOTAGE, FIGURE THE ENTIRE CROSS SECTION VOLUME INCLUDING THE OPEN PORTION OF THE FLUE AND FIREBOX. THE FOOTING SHOULD EXTEND AT LEAST BELOW THE DEPTH OF THE GREATEST FROST PENETRATION.

AS A GENERAL RULE, FOOTINGS SHOULD BE OF CONCRETE AT LEAST TWELVE INCHES THICK AND SHOULD EXTEND AT LEAST SIX INCHES ON ALL SIDES OF THE FOUNDATION. CONCRETE SHOULD BE POURED ON UNDISTURBED SOIL.

FOUNDATION WALLS SHOULD BE A MINIMUM OF EIGHT INCHES THICK.

SEE: *CHIMNEY HEIGHTS, FIREPLACE PARTS, CIRCULATING FIREPLACES.*

ALSO: BOOK OF SUCCESSFUL FIREPLACES (STRUCTURES PUBLISHING CO.)

(7)

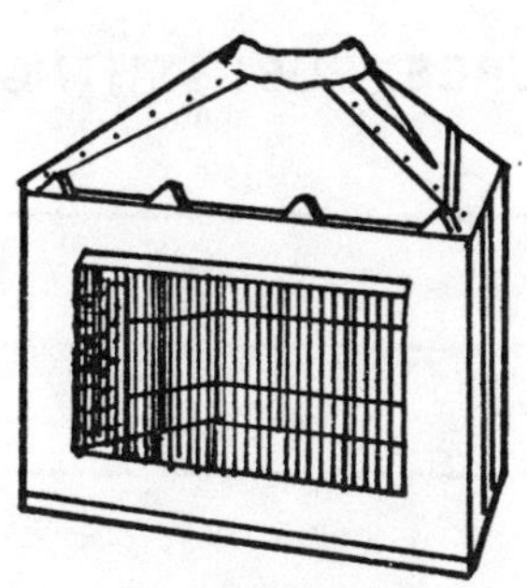

FIREPLACE UNIT

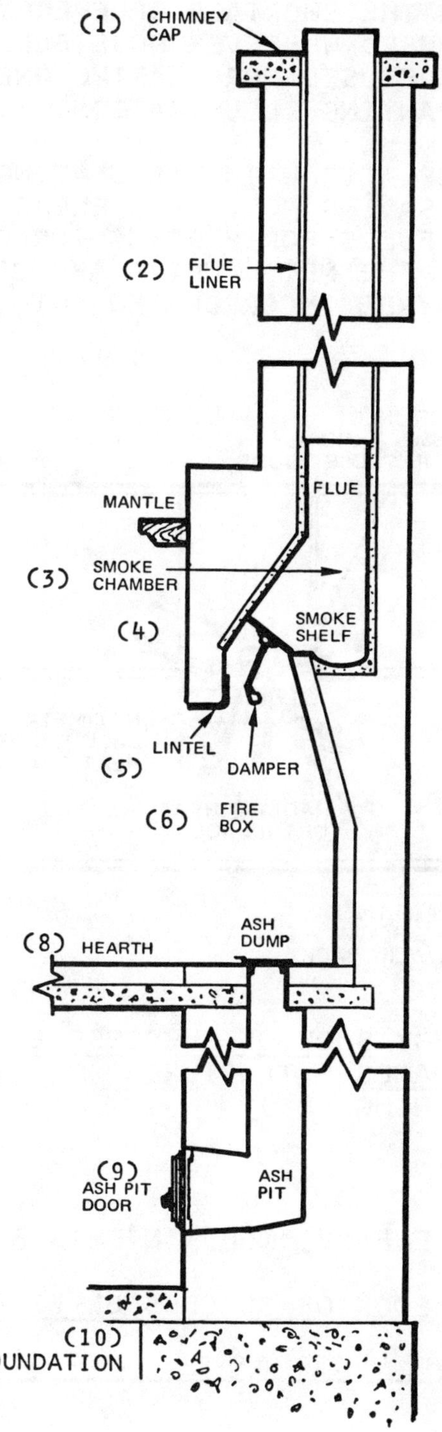

COURTESY BOOK OF SUCCESSFUL FIREPLACES (STRUCTURES PUBLISHING CO.)

CIRCULATING FIREPLACES-----

WITH THE SHORTAGE OF ENERGY, CIRCULATING FIREPLACES ARE NOW BEING USED WHEREVER POSSIBLE IN NEW HOMES AND REMODELING PROJECTS. THEY ARE USED FOR HEATING ONE OR MORE ROOMS AS INDICATED BY THE ACCOMPANYING ILLUSTRATIONS.

FIREPLACES WITH CIRCULATING UNITS ARE PRACTICAL IN EVERY WAY WHILE SAVING MONEY AND ENERGY BY UTILIZING LESS EXPENSIVE WOOD OR OTHER FUELS FOR HEATING PURPOSES. MANY TYPES AND SIZES ARE AVAILABLE. THE METAL UNITS ARE USED AS THE BASIC PART OF THE FIREPLACE AND MAY BE INCORPORATED INTO MANY STYLES AND DESIGNS.

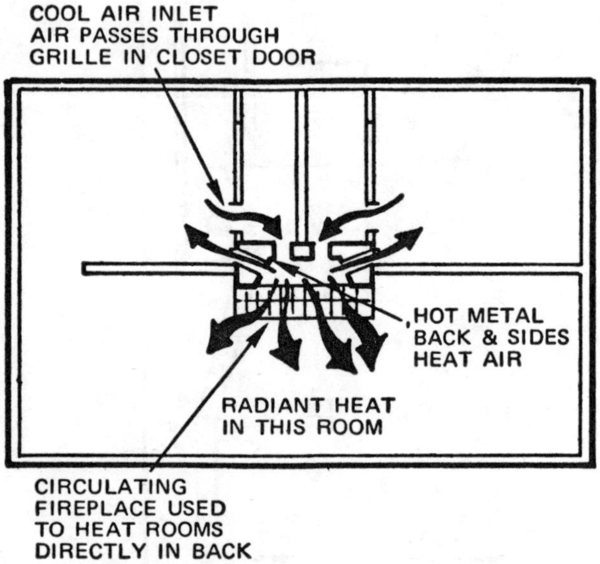

COURTESY BOOK OF SUCCESSFUL FIREPLACES (STRUCTURES PUBLISHING CO.)

COURTESY OF A. R. WOOD.

SEE: *ENERGY, WOOD* IN THIS BOOK.

ALSO: BOOK OF SUCCESSFUL FIREPLACES (STRUCTURES PUBLISHING CO.)

Notes: ______________________________

____

____

____

____

____

CLIPS AND CONNECTORS, DESCRIPTIONS, NAMES, USES-----

METAL CLIPS AND WOOD CONNECTORS ARE MADE IN SEVERAL TYPES TO SUIT MOST CONDITIONS FOUND IN NORMAL LIGHT CONSTRUCTION WORK. HERE ARE THE NAMES, USES AND DESCRIPTION OF THE VARIOUS TYPES USUALLY AVAILABLE. THE DETAILS AND USES FOR EACH TYPE LISTED ARE SHOWN. THE FIGURE NUMBER IN THE DESCRIPTION REFERS TO THE SKETCH NUMBER.

ANCHOR CLIPS — USED TO ANCHOR VARIOUS SIZES AND TYPES OF WOOD TO CONCRETE MASONRY. FIGURE 1

BEAM CLIPS — TO ANCHOR THE TOP OF A WOOD POST OR COLUMN TO A WOOD BEAM. FIGURE 3

BRIDGING CLIPS — FOR USE BETWEEN FLOOR OR CEILING JOISTS AS BRIDGING. FOR 2 X 8 OR 2 X 10 ON 16" CENTERS. FIGURE 4

DRYWALL CLIPS — TO INSTALL DRYWALL. ELIMINATES CORNER STUDS AND BACKUP STRIPS. FIGURES 5 AND 8

FRAMING CLIPS — FOR UNIVERSAL RIGHT OR LEFT HAND TWO AND THREE-WAY TIES BETWEEN FRAMING MEMBERS. FIGURE 6

JAMB CLIPS — SELF-NAILING CLIPS FOR INSTALLING PRE-HUNG DOORS. ELIMINATES SHIMMING AND NAIL MARKS ON THE FACE OF THE JAMB. FIGURE 7

JOIST CLIPS — FOR ATTACHMENT OF FLOOR AND CEILING JOISTS TO THE HEADERS. SINGLE AND DOUBLE TYPES FOR SIZES FROM 2 X 4 OR 6, TO 2 X 14. FIGURE 12

PANEL CLIPS — USUALLY USED ON PLYWOOD ROOF SHEATHING. AVAILABLE IN STANDARD PLYWOOD THICKNESSES. FIGURE 11

PLATE TIES — TO CONNECT WALLS AND PARTITIONS. FIGURE 9

POST BASE CLIPS — FOR FAST ADJUSTABLE INSTALLATION OF 4" X 4" WOOD POSTS TO CONCRETE. FIGURE 2

PURLIN CLIPS — TO PLACE PURLINS BETWEEN ROOF TRUSSES OR RAFTERS. SADDLE AND SINGLE TYPES ARE USED. FIGURE 10

STORM CLIPS — TO ANCHOR ROOF TRUSSES OR RAFTERS TO PLATES AND STUDS TO GUARD AGAINST WIND DAMAGE. FIGURE 13

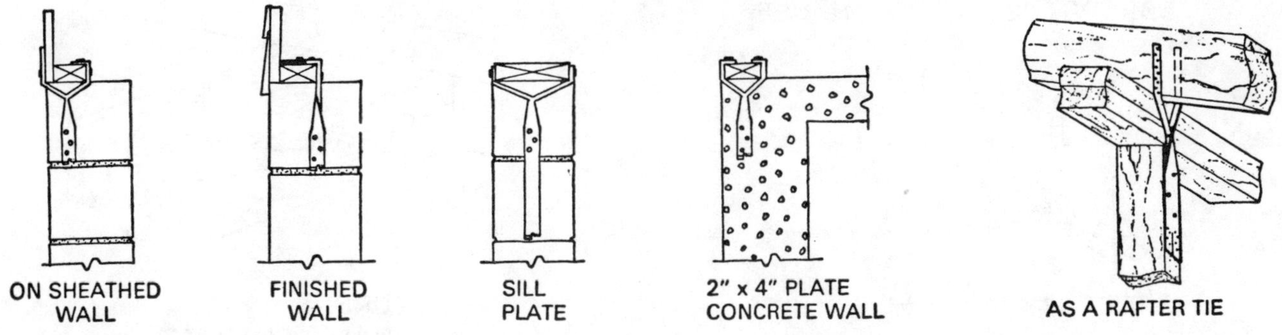

FIGURE 1
USES FOR ANCHOR CLIPS

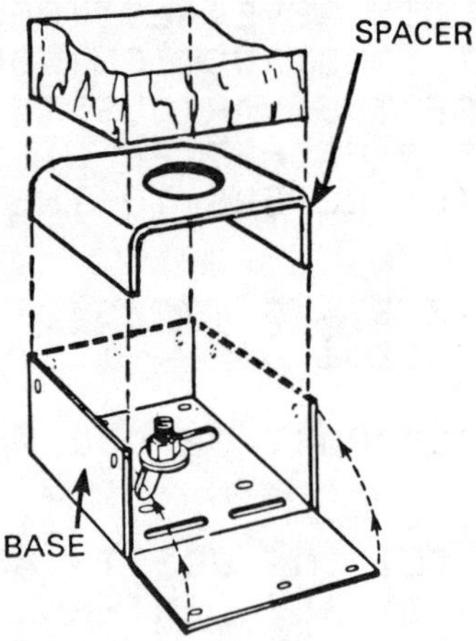

FIGURE 2
POST BASE CLIP IN USE

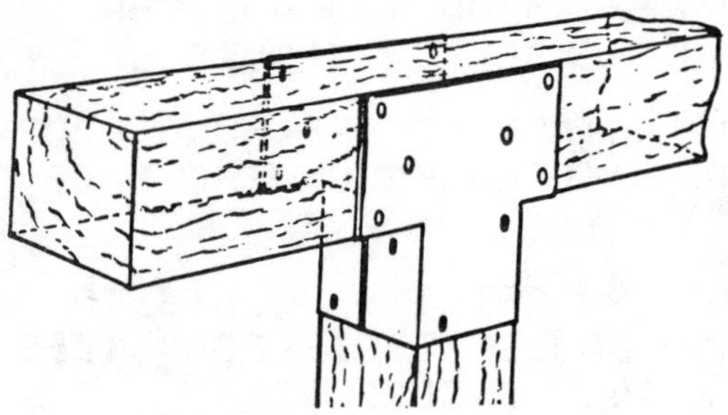

FIGURE 3
A BEAM CLIP IN USE

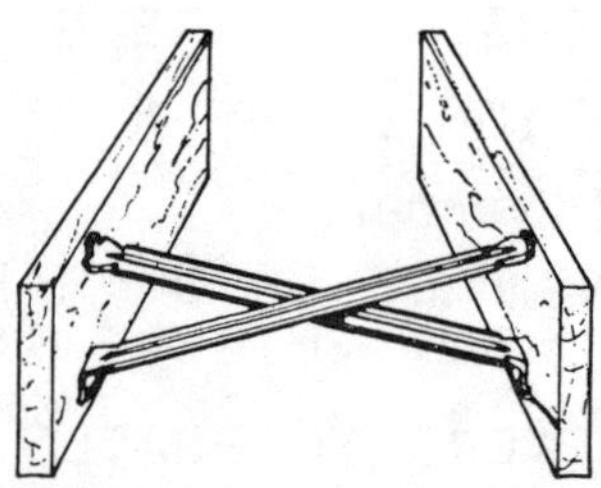

FIGURE 4
BRIDGING CLIPS IN USE

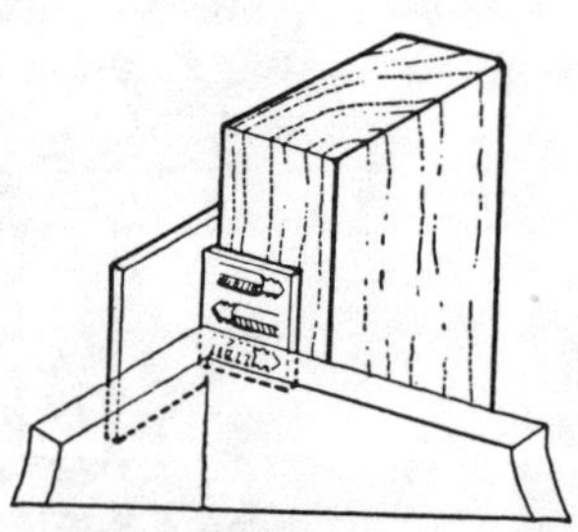

FIGURE 5
DRYWALL CLIP CORNER

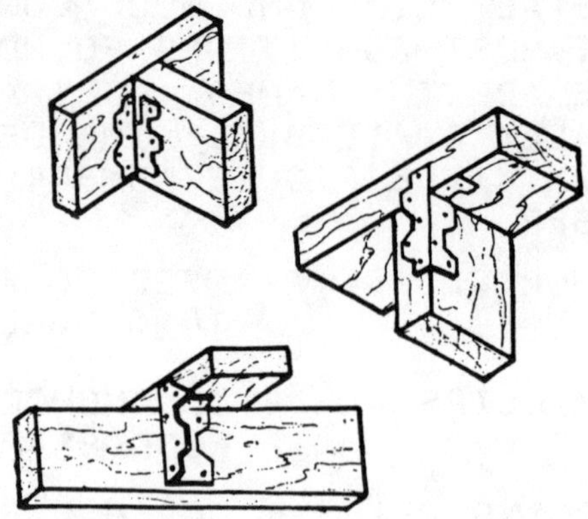

FIGURE 6
USE OF FRAMING CLIPS

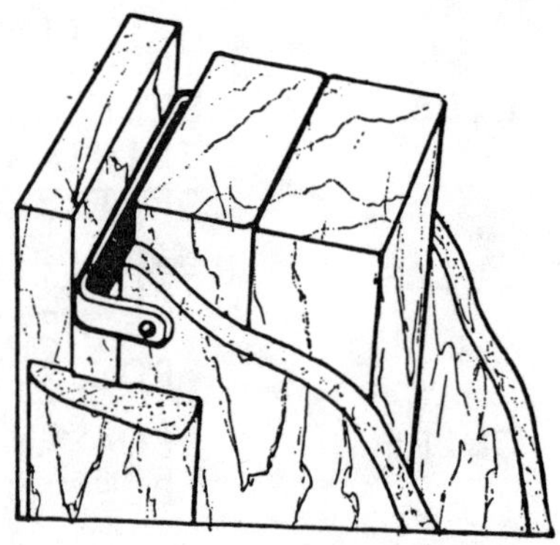

FIGURE 7
HIDDEN JAMB CLIP

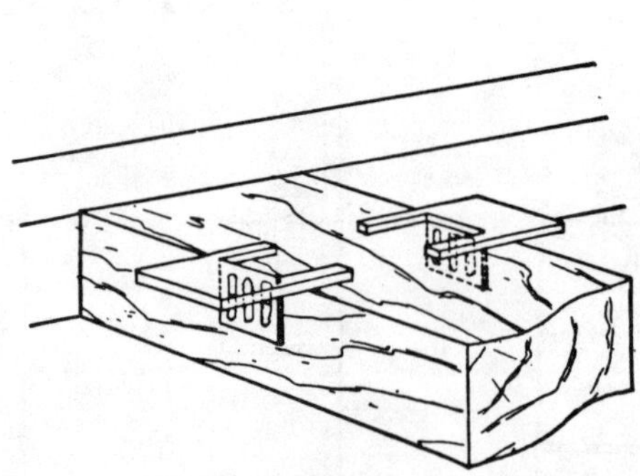

FIGURE 8
DRYWALL CLIP AT
CEILING CORNER

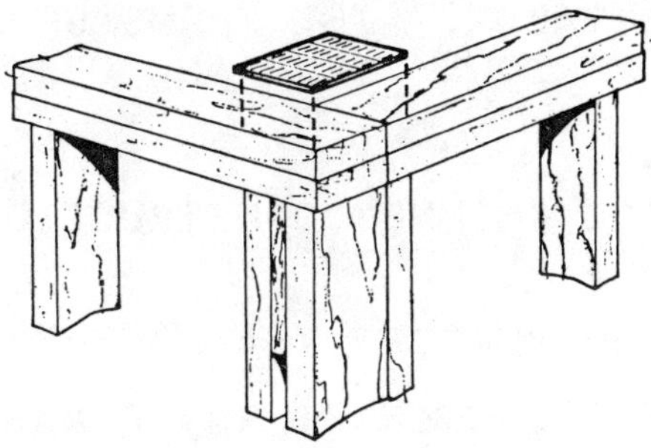

FIGURE 9
USE OF TOP PLATE TIE

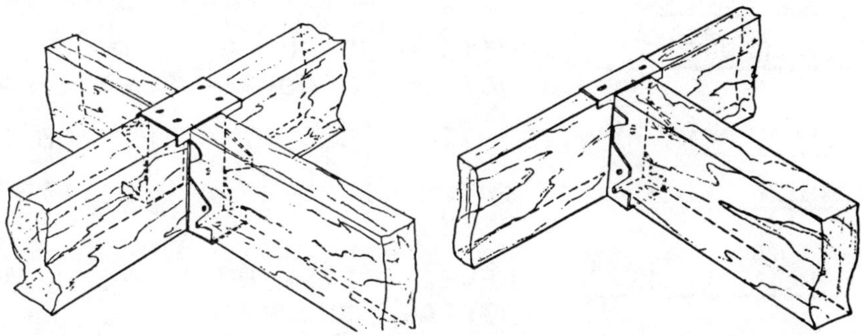

FIGURE 10
PURLIN CLIPS IN PLACE

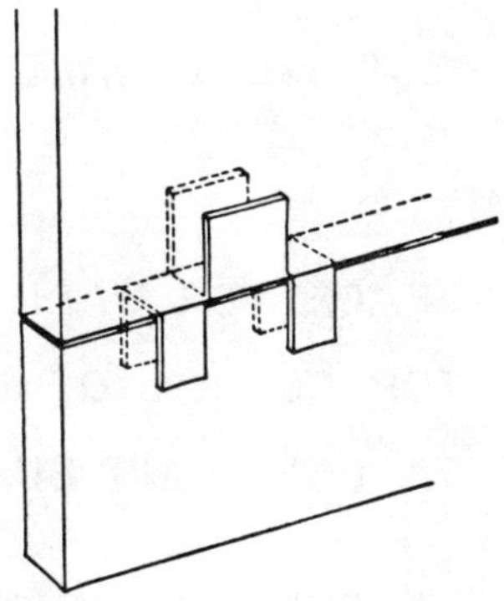

FIGURE 11
PANEL CLIPS IN USE

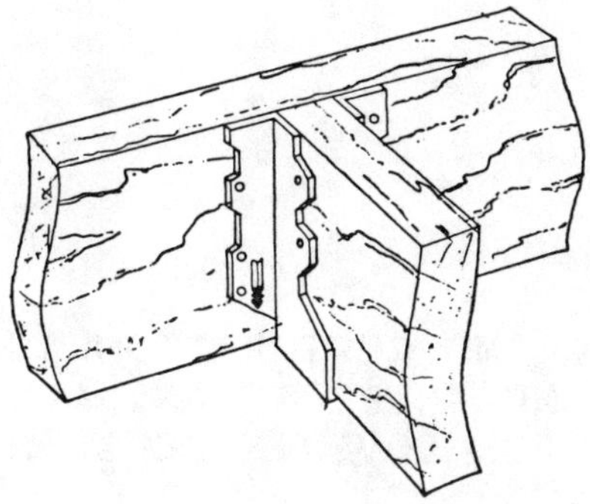

FIGURE 12
JOIST CLIP IN USE

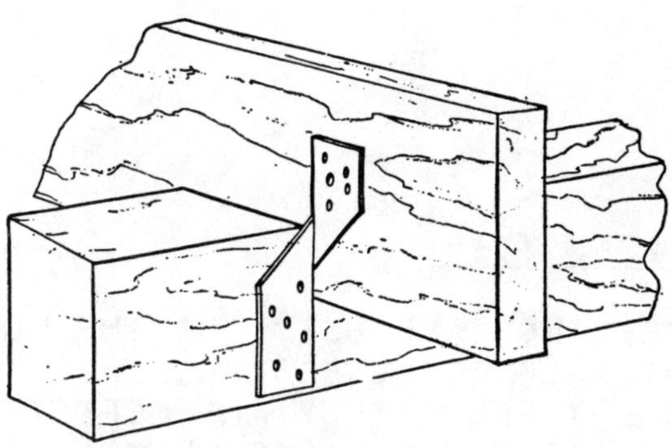

FIGURE 13
STORM CLIP IN USE

COURTESY <u>BUILD IT WITH CLIPS</u> (THE PANEL CLIP CO.)

CONCRETE MASONRY, TERMS-----

CONCRETE MASONRY UNITS ARE USED UNIVERSALLY. THE TERMS SHOWN HERE WILL ENABLE YOU TO UNDERSTAND THE VARIOUS TYPES AND USES.

CHIMNEY BLOCK. SQUARE UNITS DESIGNED FOR CHIMNEY CONSTRUCTION. (USUALLY WITH FLUE LINERS.)

CONCRETE BRICK. SOLID UNITS. USUALLY NOT LARGER THAN 4"X4"X12".

DOUBLE CORNERS. UNITS WITH FLAT ENDS FOR USE IN PIERS, PILASTERS OR CORNERS.

HOLLOW BLOCK. ALSO KNOWN AS STRETCHERS.

PARTITION BLOCK. 4" OR 6" THICK UNITS USED FOR PARTITIONS.

PATIO BLOCK. THIN SOLID UNITS FOR USE IN WALKS, PATIOS, TERRACES.

PILASTER BLOCK. UNITS USED FOR BUILDING PILASTERS. ALSO USED ON EDGE FOR DECORATIVE WALL PATTERNS.

SASH BLOCK. UNITS WITH A SLOT IN THE JAMB FOR METAL SASH.

SCREEN BLOCK. 4" THICK, 12"X12" AND 16"X16" FOR DECORATIVE PURPOSES.

SHADOWALL BLOCK. BEVELED FACE UNITS TO PROVIDE UNUSUAL WALL APPEARANCE.

SILL BLOCK. SOLID UNITS FOR SILL CONSTRUCTION.

SLUMP BLOCK. UNITS WITH A SAGGING OR SLUMPING FACE FOR WALL USES.

SOLID TOP. UNITS WITH A SOLID TOP FOR USE IN TOP COURSE OF WALL.

SPLIT BLOCK. UNITS WITH A SPLIT OR FRACTURED FACE FOR UNUSUAL WALL TREATMENT.

STRETCHER BLOCK. ALSO KNOWN AS HOLLOW BLOCK. (REGULAR UNITS.)

WOOD SASH JAMB BLOCK. UNITS WITH RECESS FOR SASH UNITS IN THE JAMB.

CEILING TILE-----

CEILING TILE USED FOR THE HOME, REMODELING AND RESTORATION IS AVAILABLE IN SEVERAL SURFACES AND PATTERNS AND IN PLAIN AND ACOUSTICAL TYPES. THEY ARE ESPECIALLY ADAPTABLE FOR COVERING CEILINGS IN OLDER HOMES WHERE THE PLASTER HAS FAILED.

THEY ARE MADE IN TWO SIZES—12"X12" AND 12"X24"—USUALLY 1/2" THICK. THEY MAY BE CEMENTED, STAPLED OR ATTACHED TO THE CEILING BY METAL PARTS, SOLD WITH THE TILE. FOR NORMAL USE THE TILE MADE OF PRESSED WOOD OR CANE FIBERS MAY BE USED. WHERE HIGH

CEILING TILE (continued)

FIRE RATINGS ARE DESIRED OR REQUIRED, MINERAL FIBER TYPES ARE AVAILABLE.

SUSPENDED CEILINGS USE LIGHT METAL ANGLES AND 24"X48" PANELS. THIS TYPE IS OFTEN USED IN KITCHEN OR FAMILY AREAS WHERE LIGHTED CEILING PANELS OR GRIDS ARE DESIRED. INSULATING AND ACOUSTICAL TYPES ARE AVAILABLE IN SEVERAL FINISHES. THE ACOUSTICAL TYPES CAN ABSORB UP TO 70 PERCENT OF THE EXCESS NOISE IN A ROOM.

VARIOUS PATTERNS OF THE 12"X12" SIZE ARE SHOWN HERE. COLORS VARY FROM WHITE TO LIGHT COLORED BLENDS AND TEXTURES.

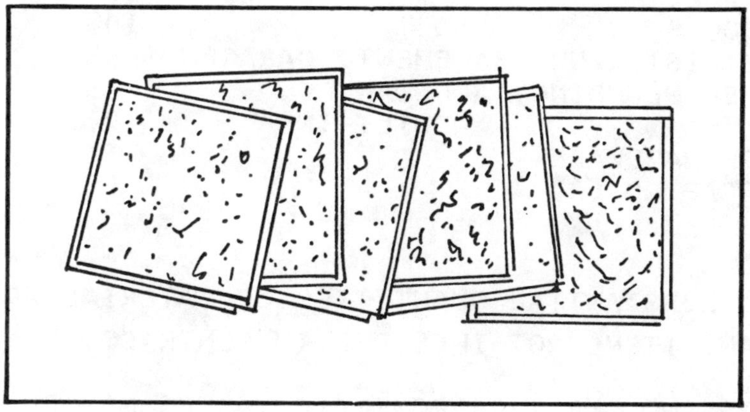

CUBING-----

DETERMINING THE CUBIC CONTENTS OF A BUILDING IS OFTEN A TRICKY PROBLEM. HERE IS A SIMPLIFIED FORMULA OR SHORT METHOD TO CALCULATE THE CUBIC VOLUME OF A HOUSE.

USE AVERAGE HEIGHTS BELOW MULTIPLIED BY OUTSIDE DIMENSIONS TO GET THE CUBIC FOOTAGE.

HEIGHTS	WITHOUT BASEMENT	WITH BASEMENT	FIRST FLOOR	SECOND FLOOR
1 STORY	2'	8'	9'	--
1-1/2 "	2'	8'	9'	6'
2 "	3'	8'	9'	9'

ADD FOR VARIOUS ROOF TYPES:

GABLE -	50% OF TOTAL ROOF VOLUME.	WIDTH X LENGTH X HEIGHT
HIP -	45% " " "	NONE
SHED		WIDTH X LENGTH X AVERAGE HEIGHT
GAMBREL		WIDTH X LENGTH X HEIGHT

CONSTRUCTION COSTS-----

THIS SIMPLIFIED CHART WILL SHOW YOU THE APPROXIMATE VALUE OF VARIOUS COMPONENTS OF HOUSE CONSTRUCTION AND THE ACCUMULATIVE VALUE AS CONSTRUCTION IS UNDERWAY.

COMPONENT	COMPONENT VALUE	ACCUMULATIVE VALUE
FOUNDATION AND ROUGH GRADING	10%	10%
ROUGH FLOOR (1ST OR SLAB)	5%	15%
ROUGH FRAMING, ROOF SHEATHING (1)	10%	25%
ROOF	5%	30%
CHIMNEY, ROUGH FIREPLACE	5%	35%
SIDING, WINDOWS, DOOR FRAMES (MAY VARY)	5%	40%
BRICK VENEER, STONE	5%	45%
ROUGH PLUMBING	5%	50%
ROUGH WIRING, ROUGH HEATING	5%	55%
PLASTER OR DRY WALL	10%	65%
FINISH, TRIM, DOORS	10%	75%
FINISHED FLOORS: 1ST, 2ND, BASEMENT, GARAGE	5%	80%
HEATING, FINISHED PLUMBING, APPLIANCES, BUILT-INS	10%	90%
LIGHTING, FINISH HARDWARE	5%	95%
PAINTING, LANDSCAPING (2)	5%	100%

NOTE:

(1) ADJUST WHEN SUBSTANTIAL QUANTITIES OF MATERIAL ARE ON THE SITE.

(2) ALLOW FOR ANY ITEMS NOT INCLUDED IN CONTRACT.

ANY COMPONENTS AND THEIR APPROXIMATE VALUES MAY BE COMBINED AS THEY ARE COMPLETED TO SHOW TOTAL VALUE OF WORK CREDIT.

Notes: ____________________

CONDOMINIUMS, TOWNHOUSES, COOPERATIVES-----

THERE IS A WIDE VARIANCE IN CONTRACTS AND AGREEMENTS COVERING SALES AND SALES/LEASE PLANS FOR THIS TYPE HOUSING. BASICALLY, A CONTRACT SHOULD PROVIDE A STIPULATED PORTION OF THE TOTAL PROPERTY WITH CLEAR TITLE TO COVER. TAXES, ASSESSMENTS, ETC. SHOULD ALSO BE BASED ON A PERCENTAGE OF THE TOTAL WITH PREDETERMINED DATA ON MAINTENANCE, UTILITIES, ETC. COVERED IN THE AGREEMENT. AN IMPORTANT FEATURE IS THE PLAN OF OVERALL MANAGEMENT WHICH MIGHT BE HANDLED BY THE BUILDERS OR OWNERS' COMMITTEE. THIS SHOULD BE SPELLED OUT FULLY IN THE CONTRACT TO SHOW HOW THE MATTER WOULD BE EXECUTED UNDER VARIOUS TYPES OF OWNERSHIP AND MANAGEMENT. RESALE INFORMATION IS ALSO VERY IMPORTANT IN PROPERTY OF THIS TYPE. THE USUAL PRECAUTIONS AND CHECKING THAT WOULD PREVAIL IN THE PURCHASE OF ANY TYPE OF REAL ESTATE SHOULD BE EXERCISED. THIS WOULD INCLUDE LEGAL ADVICE IN MOST CASES.

CONTRACTS-----

YOU SHOULD REQUIRE A CONTRACT OR AGREEMENT ON EVERY TRANSACTION COVERING THE PURCHASE OR SALE OF REAL PROPERTY, BUILDING OR REMODELING. CAREFULLY READ AND STUDY THE CONDITIONS AND TERMS BEFORE SIGNING ANY PAPERS. IN MOST CASES WHERE FINANCING IS HANDLED THROUGH A RECOGNIZED LENDING INSTITUTION OR AGENCY, COPIES OF THE CONTRACT OR AGREEMENT—AS WELL AS DEEDS, PLANS, SPECIFICATIONS, ETC.—WILL BE REQUIRED BEFORE THE LOAN IS CONSUMMATED.

CORNER CUPBOARDS-----

CORNER AND CHINA CUPBOARDS AND CLOSETS ALWAYS UTILIZE THE CORNERS TO BEST ADVANTAGE. SOME OF THE DIMENSIONS OF PRE-BUILT TYPES ARE SHOWN HERE. ON THE JOB TYPES ARE BUILT TO SUIT YOUR NEEDS.

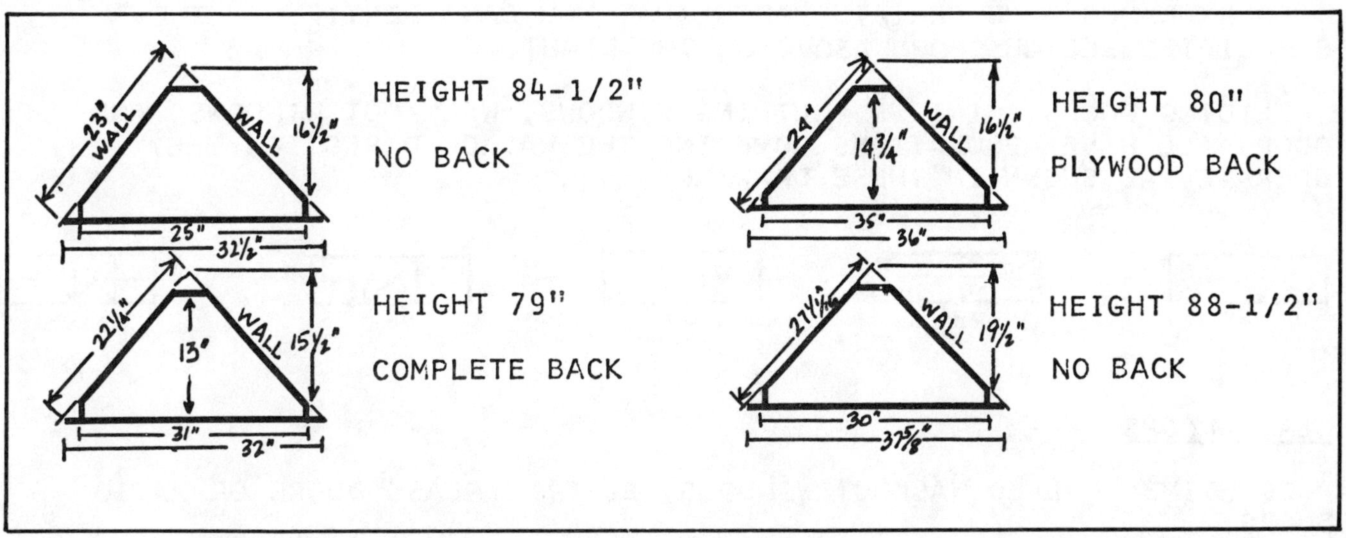

DOOR PARTS & FRAMES-----

DOORS ARE MADE IN MANY TYPES AND STYLES. THEY ALL HAVE BASIC PARTS. THE DRAWING ON THE LEFT SHOWS THE NAME AND LOCATION OF ALL THE PARTS.

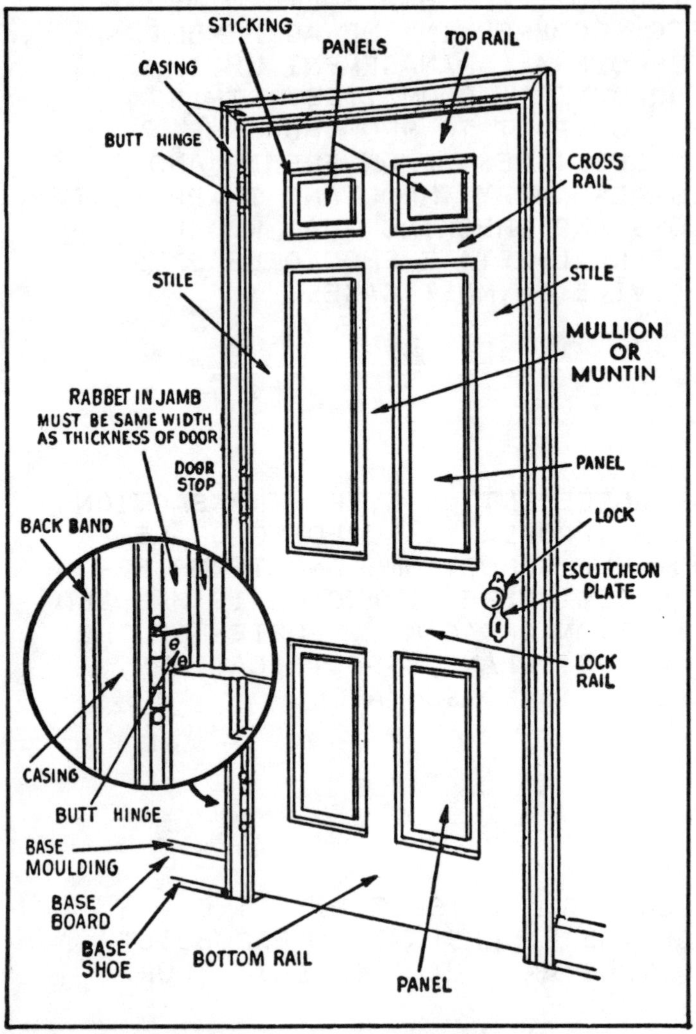

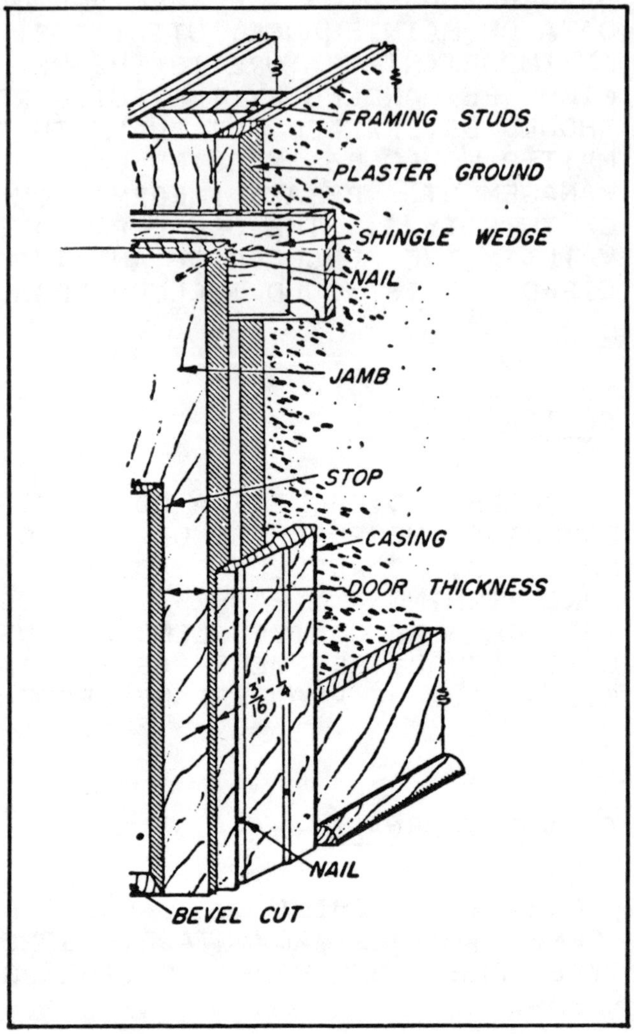

DOOR FRAMES HAVE CERTAIN PARTS THAT YOU CAN IDENTIFY IN THE SIMPLIFIED SKETCH SHOWN ABOVE ON THE RIGHT.

SLIDING DOORS, WINDOWS, GLIDING WINDOWS, WALK-OUT WINDOWS, AND DOORS ALL HAVE REGULATIONS COVERING THE WAY OR DIRECTION THEY OPERATE, AS SHOWN IN THESE DRAWINGS.

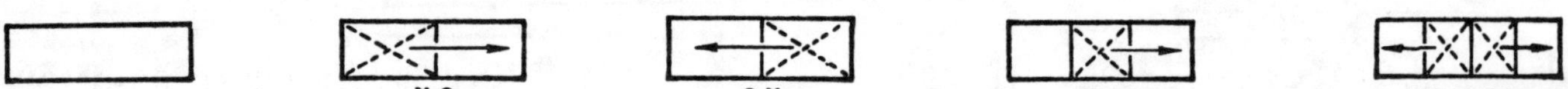

OF X-O O-X O-X-O O-X-X-O

GLASS DOORS-----

SOMETIMES CALLED WALKOUT WINDOWS, SLIDING GLASS DOORS OR PATIO DOORS.

DOOR TYPES-----

NORMALLY TWO KINDS OF STANDARD-SIZED DOORS ARE AVAILABLE FOR USE IN HOUSE CONSTRUCTION. THEY ARE KNOWN AS FLUSH OR FLAT DOORS AND RAISED PANEL OR DECORATIVE TYPE DOORS.

THEY ARE MANUFACTURED OF VARIOUS KINDS OF WOOD, LUMBER, LAMINATED WOOD PRODUCTS, METAL, COMPOSITION COVERED OR ALL METAL.

DOORS FOR EXTERIOR USE ARE USUALLY 1-3/4" OR 2-1/4" THICK AND INTERIOR TYPE DOORS MAY BE 1-1/8", 1-3/8" OR 1-3/4" IN THICKNESS. EXTERIOR DOORS REQUIRE TREATMENT, PAINT OR WEATHER PROTECTION AROUND ALL EXPOSED EDGES AND BOTH SIDES.

LUXURY ENTRANCE DOORS WITH CARVED OR EXOTIC WOOD DESIGNS ARE FREQUENTLY FURNISHED IN PAIRS FOR DOUBLE-DOOR INSTALLATION. EXTERIOR DOORS ARE INCREASINGLY THERMALIZED WITH SPECIAL FACTORY INCLUDED FEATURES SUCH AS INSULATIVE CORES AND MAGNETIC WEATHER-STRIPS.

SHOWN ARE SOME OF THE STANDARD TYPES, STYLES AND DESIGNS. PATIO DOORS ARE OFTEN CALLED SLIDING DOORS. NOT SHOWN HERE ARE FRENCH DOORS, MIRRORED, STORM, OR LOUVERED DOORS.

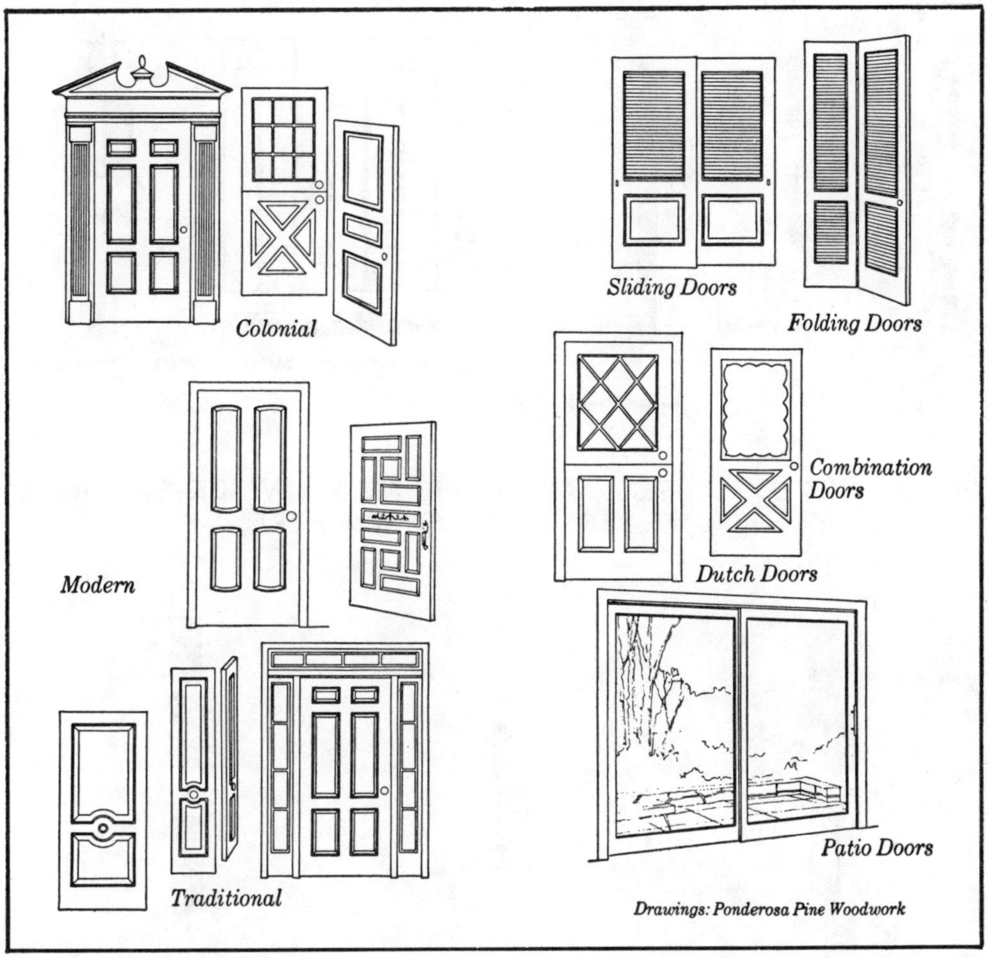

Drawings: Ponderosa Pine Woodwork

SEE: DOOR PARTS & FRAMES, DOOR HAND, CLOSET, CABINET, CUPBOARD DOORS, GLASS DOORS.

DOOR HAND-----

VIEWING A DOOR FROM THE INSWING SIDE, IT IS A RIGHT HAND DOOR WHEN THE LOCKING JAMB IS ON THE VIEWER'S RIGHT AND VICE VERSA AS INDICATED IN THE SKETCH BELOW.

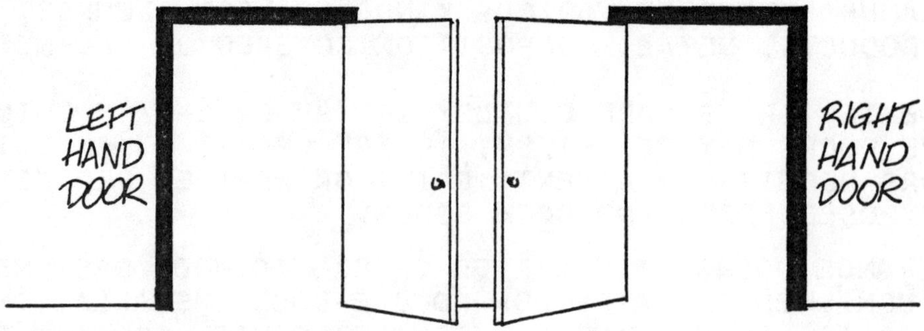

CLOSET, CABINET, CUPBOARD DOORS-----

SEVERAL DOOR STYLES AND TYPES MAY BE USED FOR THESE PURPOSES. NAMES AND TYPES ARE SHOWN BELOW.

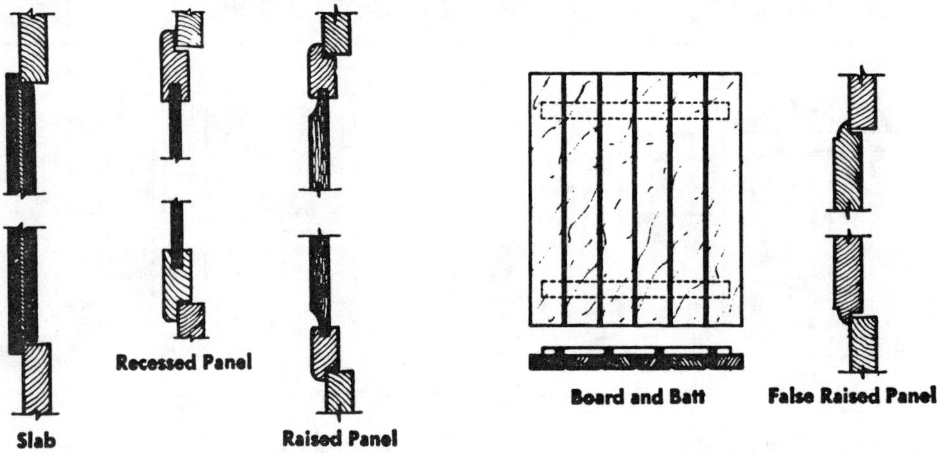

DOORS FOR CABINETS, CLOSETS AND CUPBOARDS MAY BE INSTALLED USING VARIOUS TYPES OF LIP DESIGN AS SHOWN HERE.

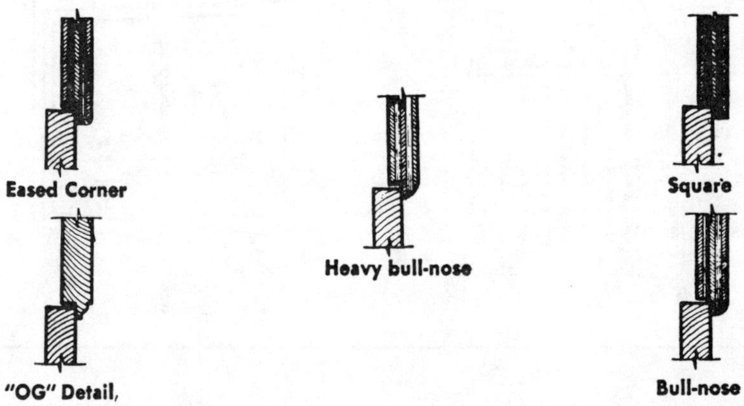

CLOSET, CABINET, CUPBOARD DOORS (continued)

THE TYPES AND TERMS USED TO INDICATE THE VARIOUS STYLES OF CABINET DOORS WILL HELP YOU IN SELECTING THE KIND YOU WANT.

LIP...STANDARD TYPE OF DOOR CONSTRUCTION INDICATED BY THE PRESENCE OF A "LIP" OR "FLANGE" EXTENDING PARTLY OVER THE FACE FRAME (OR STILES AND RAILS) OF THE CABINET.

LIP DOOR STYLES

1. SLAB
2. RAISED PANEL
3. FALSE RAISED PANEL
4. RECESSED PANEL
5. BOARD AND BATT

TREATMENTS OF "LIP" ON LIP DOOR

1. SQUARE LIP
2. BULL-NOSE LIP
3. HEAVY BULL-NOSE LIP
4. EASED CORNER LIP
5. "OG" DETAIL LIP

OVERLAY...CABINET DOOR WHICH COMPLETELY COVERS THE CABINET FACE FRAME (STILES AND RAILS). THE OVERLAY DOES NOT HAVE A "LIP," AS THE ENTIRE THICKNESS OF THE DOOR IS OUTSIDE THE CABINET FRAMEWORK.

OVERLAY DOOR STYLES

1. SLAB
2. SLAB WITH HIDDEN PULLS
3. RAISED PANEL
4. FALSE RAISED PANEL
5. RECESSED PANEL
6. BOARD AND BATT

Notes:

DRIVEWAYS-----

THE SIZE AND TYPE OF DRIVEWAYS DEPEND ON THE NUMBER OF CARS TO BE STORED AND PARKED, AND THE PROPERTY LAYOUT. THIS TABLE WILL ENABLE YOU TO APPROXIMATE THE SPACE YOU WILL REQUIRE.

DRIVEWAY

TYPE	SINGLE WIDTH	DOUBLE WIDTH
STRAIGHT	8' TO 10'	16' TO 20'
CURVED	10' TO 20'	18' TO 24'

PARKING SPACE

STANDARD CAR	9'X 8' TO 10'X20'
COMPACT OR 2 STANDARD CARS	9'X 6' TO 10'X 8'
PARALLEL	9'X34' TO 10'X50'

THE SKETCHES SHOWN HERE WILL GIVE YOU SOME IDEAS OF DRIVEWAY AND GARAGE OR CARPORT LOCATION AND THE APPROXIMATE SIZES TO CONSIDER. WITH SMALLER AUTOS, DIMENSIONS MAY BE ALTERED TO SUIT THE SIZE OF LOTS AND CARS TO BE USED.

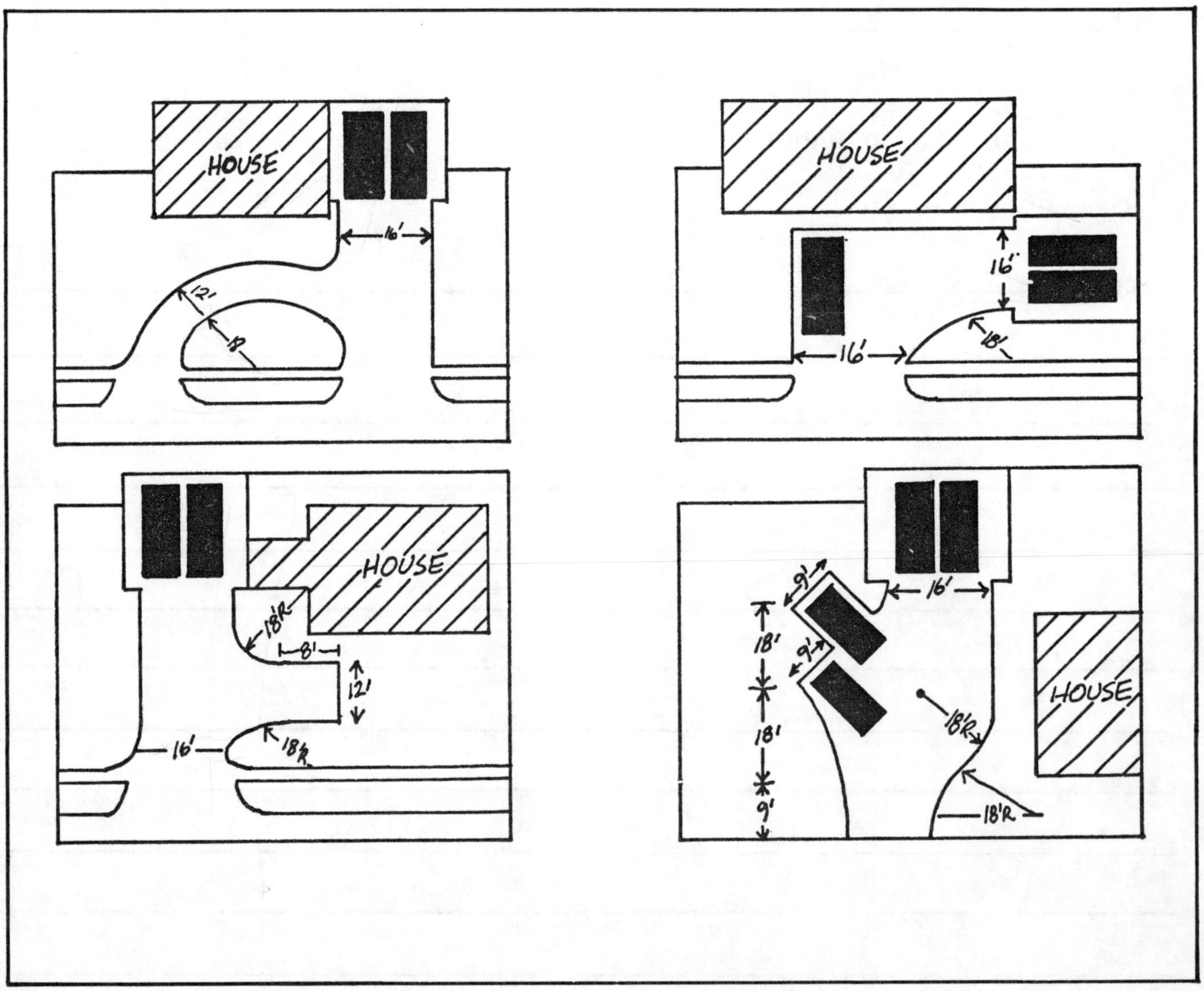

DRIVEWAYS (continued)

THE DRIVEWAY SPACE YOU WILL REQUIRE FOR CURVES AND TURNING RADIUS WILL VARY WITH CARS OF DIFFERENT WHEEL BASES. THIS CHART WILL GIVE YOU SOME IDEAS TO SUIT YOUR CONDITIONS.

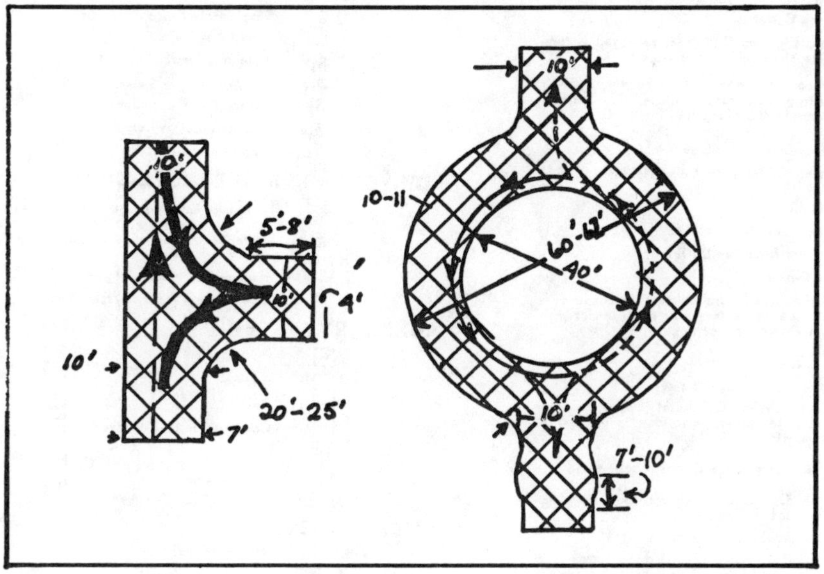

Notes: ______________________________

ELECTRICAL SYMBOLS-----

THIS CHART WILL ENABLE YOU TO UNDERSTAND THE MANY ELECTRICAL SYMBOLS FOUND THROUGHOUT HOUSE PLANS.

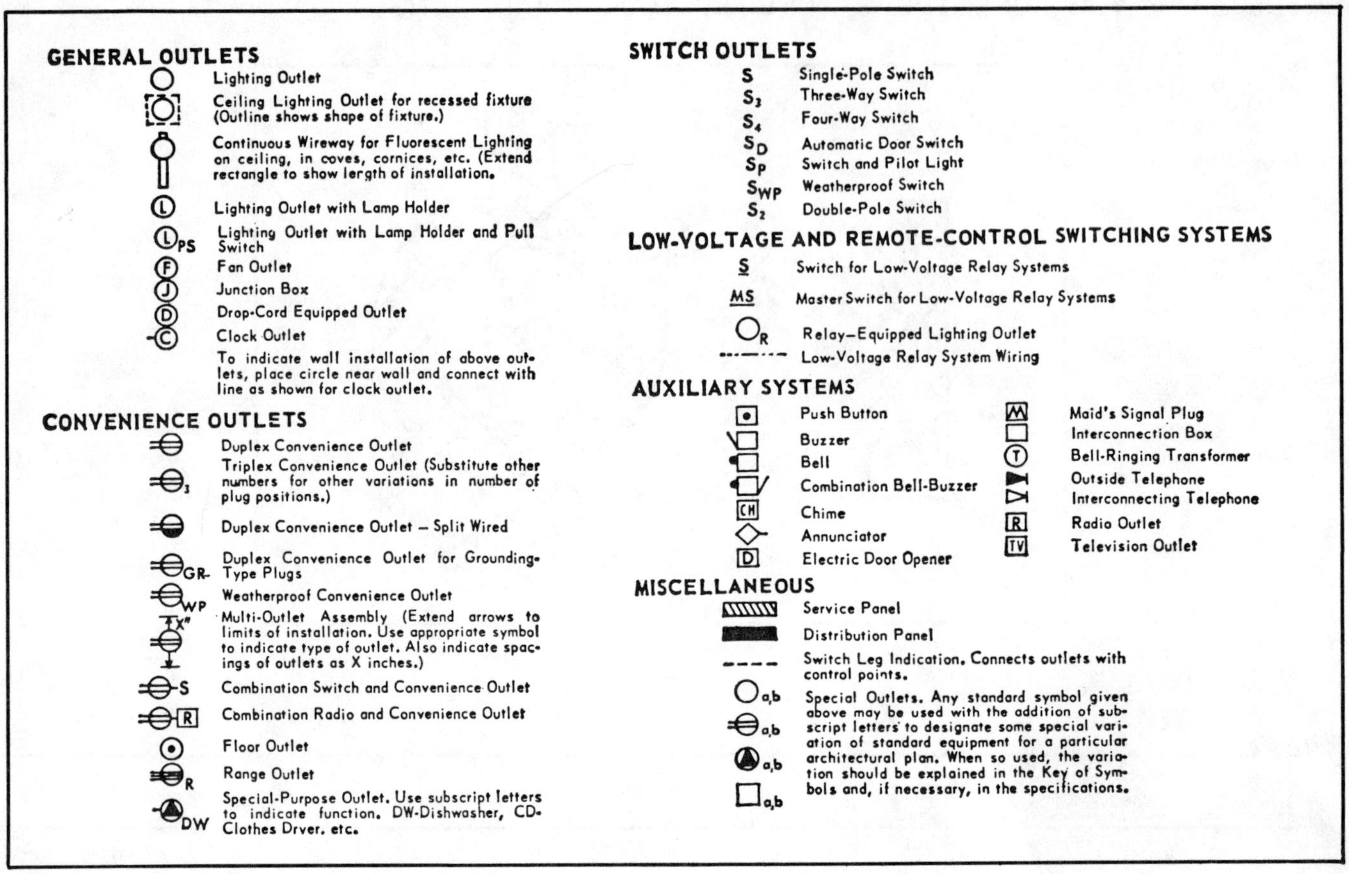

COURTESY INDUSTRY COMM. ON INTERIOR WIRING DESIGN.

Notes:

ELECTRICAL APPLIANCES AND WATTAGE REQUIREMENTS-----

THIS LIST OF APPLIANCES SHOWS THE APPROXIMATE WATTAGE REQUIREMENTS UNDER NORMAL USAGE.

APPLIANCE	TYPICAL WATTAGE
AIR CONDITIONER	1100
ATTIC FAN	400
AUTOMATIC TOASTER	1200
AUTOMATIC WASHER	700
BROILER	1000
BUILT-IN VENTILATING FAN	400
CLOTHES DRYER	4500
COFFEE MAKER	1000
DEEP FRYER	1320
DEHUMIDIFIER	350
DISHWASHER-DISPOSER	1500
DRY IRON OR STEAM IRON	1000
EGG COOKER	600
ELECTRIC BLANKET	200
FREEZER	350
FLUORESCENT LIGHTS (1 TUBE)	15-40
GRIDDLE	1000
HAIR DRYER	100
HEAT OR SUN LAMP	300
HOT PLATE	1500
IRONER	1650
LAMPS, EACH BULB	40-100
MIXER	100
OIL BURNER	250
PORTABLE FAN	100
PORTABLE HEATER	1650
RADIO	100
RANGES, ELECTRIC	8000
REFRIGERATOR	200
ROOM COOLER	600
ROTISSERIE	1380
ROASTER	1380
SANDWICH GRILL	1320
TV, BLACK AND WHITE	350
VACUUM CLEANER	300
WAFFLE IRON	1320
WASTE DISPOSER	500
WATER HEATER	2500
WATER PUMP	700

SEE: *APPLIANCES, OPERATING COSTS.*

ELEVATIONS-----

THE ELEVATIONS IN A SET OF PLANS, AS REQUIRED BY MOST BUILDING DEPARTMENTS, CONSIST OF THE VIEWS FROM THE FOUR SIDES. TYPICAL FRONT, REAR AND SIDE ELEVATIONS ARE SHOWN HERE.

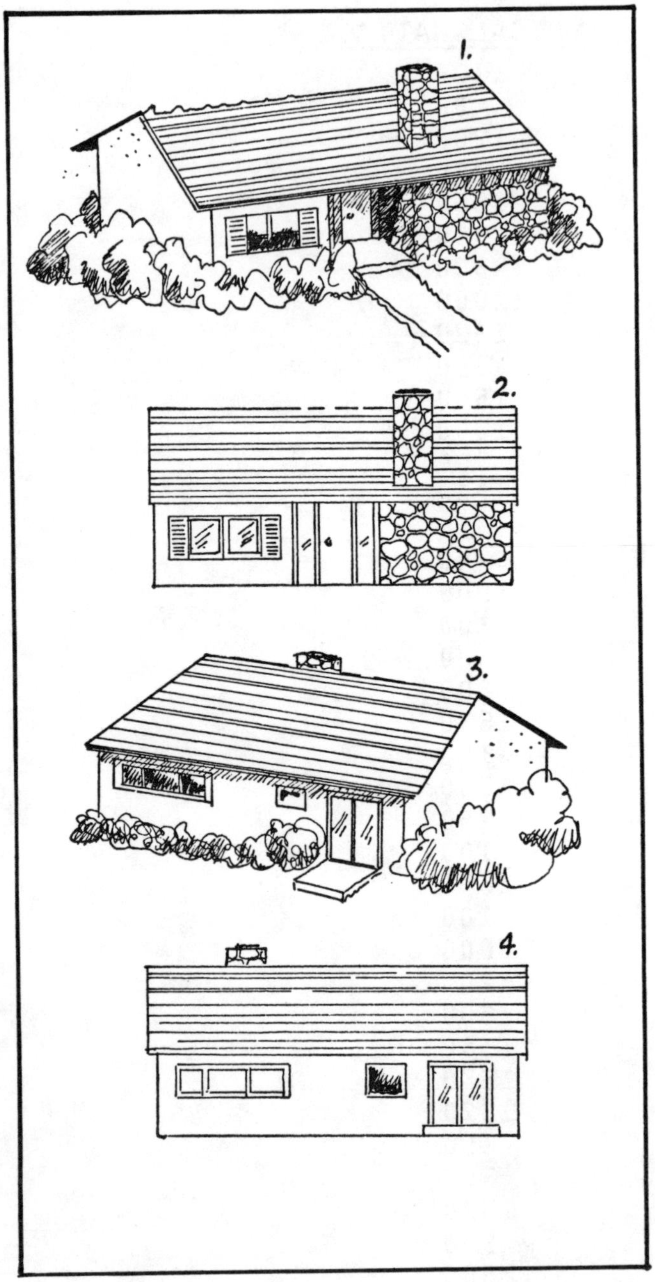

HOUSE VIEWS AND ELEVATIONS:

1. FRONT VIEW
2. FRONT ELEVATION
3. REAR VIEW
4. REAR ELEVATION
5. RIGHT SIDE VIEW
6. RIGHT SIDE ELEVATION
7. LEFT SIDE VIEW
8. LEFT SIDE ELEVATION

SEE: *HOUSE PARTS, FLOOR PLANS.*

ENERGY-----

THE USE OF HEAT OR ENERGY IN THE HOME IS MEASURED BY BRITISH THERMAL UNITS (BTU'S). WHILE THIS DATA AND INFORMATION MAY SEEM VERY TECHNICAL AND HARD TO UNDERSTAND, IT IS VERY IMPORTANT NOW THAT ALL TYPES OF ENERGY ARE BECOMING MORE EXPENSIVE FOR USE IN THE HOME.

BTU - BRITISH THERMAL UNIT. THE AMOUNT OF HEAT REQUIRED TO RAISE THE TEMPERATURE OF ONE POUND OF WATER ONE DEGREE FAHRENHEIT.

R - THERMAL RESISTANCE. A MEASURE OF ABILITY TO RETARD HEAT FLOW. R IS USED IN COMBINATION WITH NUMERALS TO DESIGNATE THERMAL RESISTANCE VALUES; FOR EXAMPLE, R-11 EQUALS 11 RESISTANCE UNITS. THE HIGHER THE R VALUE, THE HIGHER THE INSULATING VALUE. ALL INSULATION PRODUCTS HAVING THE SAME R VALUE, REGARDLESS OF MATERIAL OR THICKNESS, ARE EQUAL IN INSULATING VALUE.

U - OVERALL COEFFICIENT OF HEAT TRANSMISSION. THE COMBINED THERMAL VALUE OF ALL OF THE MATERIALS IN A BUILDING SECTION, AIR SPACES, AND SURFACE AIR FILMS. U IS EXPRESSED IN BTU'S PER HOUR PER SQUARE FOOT OF AREA PER DEGREE FAHRENHEIT OF TEMPERATURE DIFFERENCE BETWEEN INSIDE AND OUTSIDE. THE LOWER THE U, THE HIGHER THE INSULATING VALUE. R IS THE NUMERICAL RECIPROCAL OF U SO THAT $R = 1/U$. IF A WALL'S VARIOUS MATERIALS ADD UP TO A TOTAL, FOR EXAMPLE, OF 14.3 RESISTANCE UNITS, THAT WALL HAS A U VALUE OF 0.07 SINCE $14.3 = 1/0.07$.

DEEDS-----

A DEED IS A LEGAL DOCUMENT THAT IS USED TO TRANSFER OWNERSHIP OF REAL PROPERTY FROM ONE PARTY TO ANOTHER. IT SHOWS THE LEGAL DESCRIPTION OF THE LAND, SIZE, BOUNDARIES, RESTRICTIONS, ETC. DESCRIPTIONS COVER EARLY SURVEYS AND OWNERS. ALL DEEDS ARE FILED IN PROPER ORDER BY COUNTY RECORDERS, REGISTER OF DEEDS OR APPROVED OFFICIAL OFFICES AND ARE OPEN TO TITLE SEARCHERS AND THE PUBLIC.

A TRUST DEED IS LIKE A SECOND MORTGAGE AND ONLY CONVEYS SECONDARY OWNERSHIP BASED ON STIPULATED CONDITIONS.

A QUITCLAIM DEED IS A LEGAL INSTRUMENT BY WHICH SOME RIGHT, TITLE OR INTEREST IN PROPERTY IS RELEASED TO ANOTHER. IT IS A SIMPLE, EFFECTIVE WAY TO MAKE A GRANT OF PROPERTY AND IS USED TO RELEASE OR RELINQUISH ALL CLAIMS TO PROPERTY.

ELECTRICAL TERMS-----

ANY BUILDING, REMODELING OR RESTORATION YOU CONSIDER WILL ENTAIL WIRING OR CHANGES IN THE ELECTRICAL SYSTEM. THE TERMS LISTED HERE WILL HELP YOU UNDERSTAND AND DISCUSS THIS COMPLICATED PART OF BUILDING.

- ALTERNATING CURRENT. THE TYPE OF POWER USED IN MANY FARM AND HOME POWER SYSTEMS. USUALLY 60 CYCLES.
- AMPERE. THE UNIT USED IN MEASURING ELECTRICAL RATE OF FLOW.
- CIRCUIT. TWO OR MORE WIRES THROUGH WHICH ELECTRICITY FLOWS FROM THE SOURCE OF SUPPLY TO ONE OR MORE OUTLETS.
- CIRCUIT-BREAKER. BREAKS THE FLOW OF CURRENT WHEN A LINE IS OVERLOADED.
- COLOR CODE. IDENTIFICATION OF WIRES BY COLOR.
- CONDUCTORS. THE COMMON NAME FOR ELECTRICAL WIRES.

FUSE. PERFORMS THE SAME AS A CIRCUIT BREAKER IN A POWER LINE.

- GROUNDING. CONNECTING THE ELECTRICAL SYSTEM OR CIRCUIT TO THE EARTH.
- HORSE-POWER. COMMONLY USED TO INDICATE MOTOR STRENGTH. ONE HP EQUALS 746 WATTS.
- "HOT" WIRES. THE POWER-CARRYING WIRES (USUALLY BLACK OR RED) AS DISTINGUISHED FROM THE "NEUTRAL" WIRES (USUALLY WHITE).
- INSULATION. A PROTECTIVE SHEATHING USED OVER WIRES.
- OUTLET. A DEVICE THAT PERMITS TAPPING OFF ELECTRICITY AT CONVENIENT LOCATIONS.
- RECEPTACLE. A TYPE OF OUTLET INTO WHICH ELECTRIC CORDS CAN CONVENIENTLY BE PLUGGED.
- SERVICE ENTRANCE SWITCH. THE MAIN PANEL (OR FUSE CABINET) THROUGH WHICH ELECTRICITY IS BROUGHT INTO THE BUILDING AND THEN DISTRIBUTED TO VARIOUS BRANCH CIRCUITS.
- SHORT CIRCUIT. AN IMPROPER CONNECTION BETWEEN "HOT" WIRES OR BETWEEN A "HOT" WIRE AND A "NEUTRAL."
- SWITCH. A DEVICE FOR BREAKING THE FLOW OF CURRENT.
- SWITCH (3-WAY). A TYPE USED IN PAIRS TO CONTROL THE SAME LIGHT FROM DIFFERENT POINTS.
- UNDERWRITERS' LABORATORIES. A NATIONALLY ACCEPTED ORGANIZATION WHICH TESTS ALL TYPES OF WIRING MATERIALS AND DEVICES TO MAKE CERTAIN THEY MEET MINIMUM STANDARDS FOR SAFETY AND QUALITY.
- VOLT. THE UNIT USED IN MEASURING ELECTRICAL PRESSURE.
- VOLTAGE DROP. A TERM USED TO INDICATE THE VOLTAGE LOSS WHICH OCCURS WHEN WIRES ARE OVERLOADED.
- WATT. THE UNIT WHICH SHOWS CURRENT DRAIN WITH BOTH VOLTAGE AND AMPERAGE CONSIDERED.

- WATT. (CONTINUED FROM PREVIOUS PAGE)
  EXAMPLE: 1 AMPERE AT PRESSURE OF 1 VOLT IS 1 WATT.
  1 WATT USED FOR 1 HOUR IS 1 WATT HOUR.
  1000 WATT HOURS IS 1 KILOWATT HOUR (KWHR) UNIT OF MEASURE.

EASEMENTS-----

IN CASES WHERE UTILITY LINES, LEASES FOR NATURAL RESOURCES, DRIVES OR OTHER ENCROACHMENTS ON OR THROUGH PROPERTY ARE PREVALENT, THE EASEMENTS FOR THE PURPOSE ARE INCLUDED IN THE DEED OR PROPERTY DESCRIPTION. THESE ITEMS MAY BE CHECKED ALONG WITH THE TITLE SEARCH. EASEMENTS MAY BE CHECKED FURTHER ON MAPS FOUND IN LOCAL ENGINEERING OFFICES.

ENVIRONMENT AND YOUR HOME-----

WHEN SELECTING A HOMESITE, BUYING OR BUILDING A HOME, MANY THINGS CONCERNING THE ENVIRONMENT REQUIRE CAREFUL AND THOUGHTFUL CONSIDERATION. CONSIDER SCHOOLS, CHURCHES, MEDICAL FACILITIES, TRANSPORTATION, AVAILABILITY TO HIGHWAYS, EXPRESSWAYS AND PLACE OF EMPLOYMENT. IN VIEW OF THE CRITICAL ENERGY SITUATION, FUEL FOR TRANSPORTATION, AUTO USAGE AND MILEAGE ALL BECOME IMPORTANT FACTORS IN THE LOCATION OF YOUR HOME. WHILE THESE ITEMS MAY NOT DIRECTLY AFFECT YOU NOW, THEY MAY CHANGE THE VALUE OF THE PROPERTY AT THE TIME YOU SELL TO ANOTHER PERSON. INVESTIGATE THE ZONING LAWS AND DEED RESTRICTIONS COVERING THE TYPES AND SIZES OF HOMES AND BUILDINGS WHICH MAY AFFECT THE AREA IN THE FUTURE. IN OUR RAPIDLY CHANGING SOCIETY THE ENVIRONMENTAL CONDITIONS OF AN AREA CAN CHANGE VERY QUICKLY. CHECK ON NEARBY TRAFFIC CONDITIONS THAT MIGHT BE DISTURBING FROM THE STANDPOINT OF DANGER, UNDUE NOISE OR POLLUTION. CHECK THE MAINTENANCE OF PRESENT HOMES, STREETS AND LAWNS. THIS WILL GIVE YOU AN INSIGHT INTO THE TYPE OF NEIGHBORHOOD YOU ARE CONSIDERING. STREET LIGHTS AND STREET PARKING ARE ALSO IMPORTANT AND SHOULD BE CHECKED. IN RURAL AREAS—AS WELL AS MUNICIPALITIES—POLICE AND FIRE PROTECTION ARE VERY IMPORTANT. INVESTIGATE THE STREETS, CURBS, GUTTERS, STORM AND SANITARY SEWERS AND THE POSSIBILITY OF DANGER FROM FLOODING DURING HEAVY RAINS AND FROM NEARBY STREAMS THAT MIGHT OVERFLOW. IT IS ALSO ADVISABLE TO CHECK ON THE WATER SUPPLY, UTILITIES, GARBAGE AND SEWAGE DISPOSAL AVAILABLE IN THE AREA. IN RURAL LOCATIONS, CHECK ALL REGULATIONS APPLICABLE TO SANITATION INCLUDING THOSE GOVERNING THE INSTALLATION OF SEPTIC TANKS AND INDIVIDUAL DOMESTIC WATER SUPPLY SYSTEMS.

ENERGY AND HOME CONSTRUCTION-----

CHANGES IN CONSTRUCTION METHODS, INSULATION, AND HEATING AND COOLING EQUIPMENT ALL TEND TO REDUCE THE AMOUNT OF FUEL REQUIRED TO HEAT OR COOL A HOME. CHANGES IN FRAMING, USING 2 X 6 STUDS ON 24" CENTERS, REDUCES THE LUMBER COST AND UTILIZES THICKER SIDE-WALL INSULATION.

EXTENSIVE USE OF GOOD VAPOR BARRIERS REDUCES AIR INFILTRATION. WEATHER STRIPPING, CAULKING AROUND ALL DOOR AND WINDOW FRAMES, CARPETS AND GOOD PADDING, INSULATED HEAT DUCTS AND PIPES, DRAPES AND WINDOW SHADES SHOULD BE ADDED. THEY ALL TEND TO CUT FUEL COST IN ALL SEASONS.

SOME FEATURES OF A LOW ENERGY PER MONTH HOME.

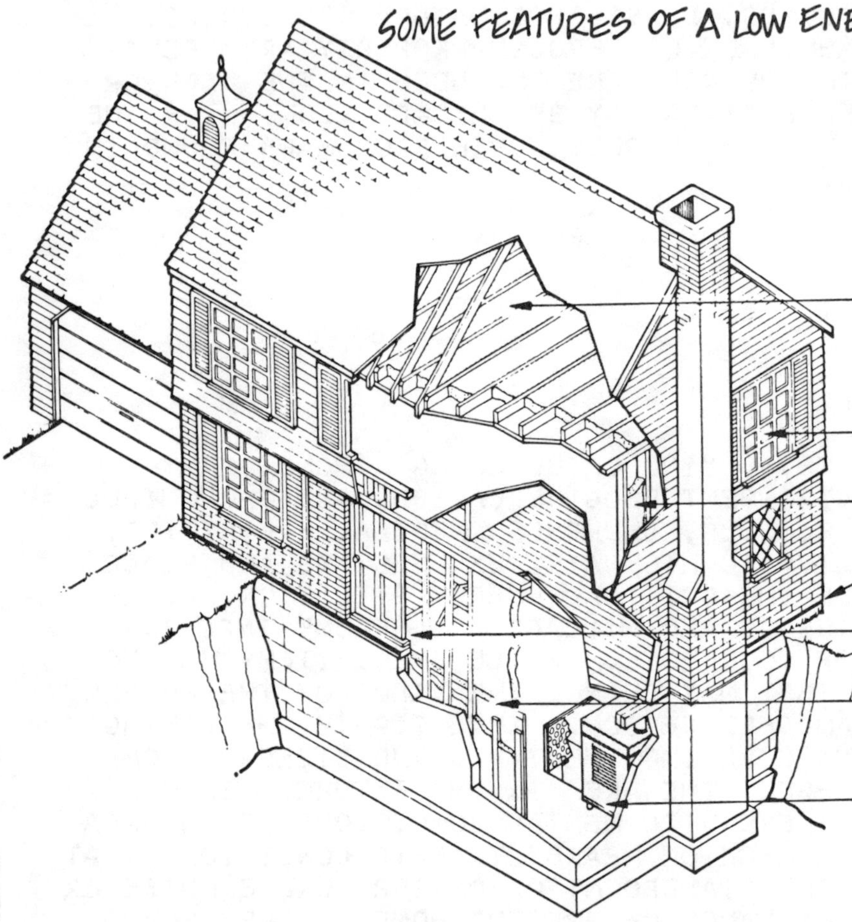

COURTESY OWENS-CORNING FIBERGLAS CORP.

SEE: *INSULATION AND YOUR HOME, INSULATION TYPES, INSULATION MATERIALS, ENERGY CONSERVATION, ENERGY EFFICIENT, "R" RESISTANCE.*

ENERGY, WOOD AS FUEL-----

THE ENERGY SHORTAGE HAS ENHANCED THE VALUE OF WOOD AS A FUEL. THIS REQUIRES THE USE OF PRACTICAL FIREPLACES AND STOVES IN MANY HOMES. THE VARIOUS KINDS AND TYPES OF WOOD USED FOR THIS PURPOSE AND THEIR PROPERTIES AND RATINGS ARE SHOWN IN THE TABLE BELOW.

HARDWOOD TREES	*Relative amount of heat*	*Easy to burn*	*Easy to split*	*Does it have heavy smoke?*	*Does it pop or throw sparks?*	*General rating and remarks*
Ash, red oak, white oak, beech, birch, hickory, hard maple, pecan, dogwood.	High	Yes	Yes	No	No	Excellent.
Soft maple, cherry, walnut	Medium	Yes	Yes	No	No	Good.
Elm, sycamore, gum	Medium	Medium	No	Medium	No	Fair—contains too much water when green.
Aspen, basswood, cottonwood, yellow-poplar	Low	Yes	Yes	Medium	No	Fair—but good for kindling.
SOFTWOOD TREES						
Southern yellow pine, Douglas-fir	High	Yes	Yes	Yes	No	Good but smoky.
Cypress, redwood	Medium	Medium	Yes	Medium	No	Fair.
White cedar, western redcedar, eastern redcedar	Medium	Yes	Yes	Medium	Yes	Good—excellent for kindling
Eastern white pine, western white pine, sugar pine, ponderosa pine, true firs.	Low	Medium	Yes	Medium	No	Fair—good kindling
Tamarack, larch	Medium	Yes	Yes	Medium	Yes	Fair.
Spruce	Low	Yes	Yes	Medium	Yes	Poor—but good for kindling.

Source: U.S. Dept. of Agriculture—Forest Service

THIS SKETCH ON THE LEFT SHOWS THE DIMENSIONS OF A CORD OF WOOD. WOOD SOLD IN SHORTER LENGTHS TO FIT MOST FIREPLACES IS KNOWN AS "FACE CORDS" OR "SHORT CORDS." THUS 24" LENGTHS WOULD BE 1/2 A CORD AND 16" LENGTHS WOULD BE 1/3 A CORD WHEN PURCHASED.

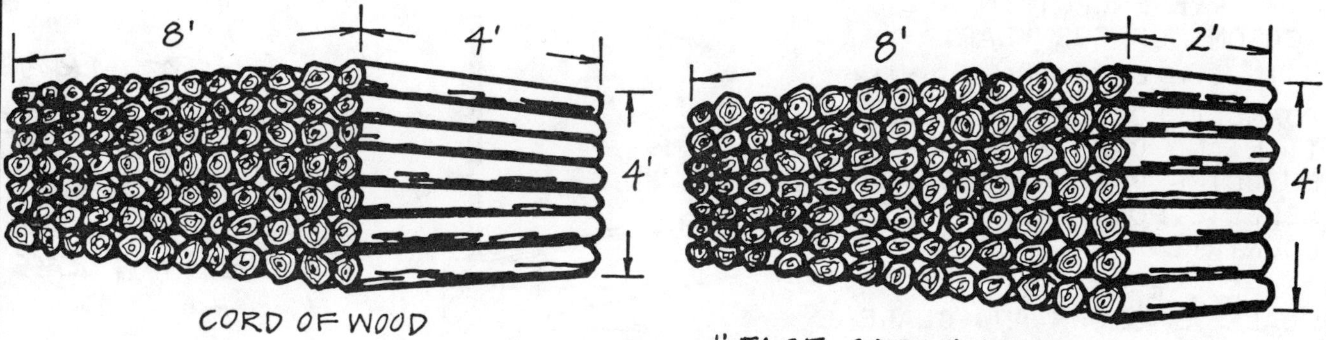

CORD OF WOOD

"FACE CORD", LOGS 24" LONG, EQUALS HALF CORD

COURTESY BOOK OF SUCCESSFUL FIREPLACES (STRUCTURES PUBLISHING CO.)

ENERGY, WOOD AS FUEL (continued)

THE HEAT DERIVED FROM A CORD OF DRY WHITE OAK WOOD IS EQUAL TO ABOUT 150 GALLONS OF FUEL OIL USED UNDER THE SAME CONDITIONS. THIS LIST OF COMMONLY KNOWN KINDS OF WOOD INDICATES THE WEIGHT AND HEAT VALUES.

Woods	Weight, lb.	Available heat, million Btu
Ash	3,440	21.0
Aspen	2,160	12.5
Beech, American	3,760	21.8
Birch, yellow	3,680	21.3
Douglas-fir	2,400	18.0
Elm, American	2,900	17.2
Hickory, shagbark	4,240	24.6
Maple, red	3,200	18.6
Maple, sugar	3,680	21.3
Oak, red	3,680	21.3
Oak, white	3,900	22.7
Pine, eastern white	2,080	13.3
Pine, southern yellow	2,600	20.5

COURTESY U.S. DEPARTMENT OF AGRICULTURE FOREST SERVICE

ENERGY CONSERVATION-----

THIS SIMPLE SKETCH WILL ENABLE YOU TO PLAN WINDOW SIZE, TYPE, EAVES OR ROOF OVERHANG TO TAKE FULL ADVANTAGE OF THE SUN AND CONSERVE ENERGY IN ALL SEASONS OF THE YEAR.

THIS EASY TO UNDERSTAND FORMULA FOR DESIGN, CONSTRUCTION AND USE OF PROPER BUILDING PRODUCTS IS ADAPTABLE TO MOST AREAS OF THE COUNTRY.

THE USE OF INSULATING GLASS AS SHOWN AND GLARE PROOF OR REFLECTIVE TYPE GLASS WILL FURTHER IMPROVE LIVING CONDITIONS AND PROMOTE ECONOMY.

SEE: *INSULATION, GLASS.*

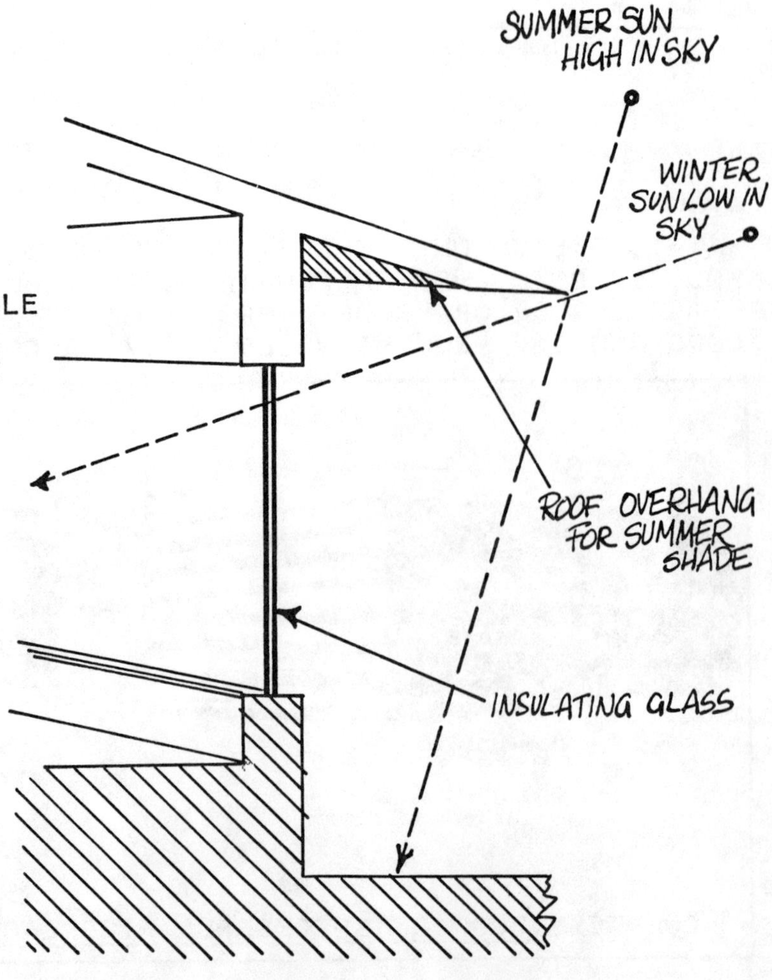

ENERGY SAVING PROGRAM-----

INCOME TAX CREDIT MAY BE AVAILABLE UNDER THE ENERGY SAVING PROGRAM FOR INSTALLING CERTAIN TYPES OF INSULATION. CHECK LOCAL IRS OFFICES FOR FURTHER INFORMATION.

ESCROW-----

WHEN A THIRD PARTY IS ASKED TO HOLD PERSONAL PROPERTY, MONEY, DEEDS, MORTGAGES, ETC., PENDING THE MEETING OF CERTAIN CONDITIONS, IT IS KNOWN AS ESCROW.

THE THIRD PARTY, OR AGENT, MAY BE A LENDING INSTITUTION, REALTOR, ATTORNEY, OR INDIVIDUAL AS CHOSEN BY THE OTHER TWO PARTIES. QUITE OFTEN IN REAL PROPERTY TRANSACTIONS AN ESCROW AGREEMENT IS USED WHEN THERE IS A QUESTION REGARDING A TITLE THAT REQUIRES CERTAIN ACTIONS OR TIME FOR PROPER CLEARANCE OR APPROVAL.

FIREPLACES, MULTIPLE OPENING-----

THESE SKETCHES SHOW THREE MULTIPLE OPENING TYPE FIREPLACES. THESE UNUSUAL TYPES ARE USEFUL, ENJOYABLE AND ENHANCE THE DECOR OF THE ROOMS THEY FACE AND SERVE.

FIREPLACES OF THIS TYPE REQUIRE CORRECT DESIGN AND CONSTRUCTION. THE FLUE SIZE AND DAMPER TYPE, AND LOCATION, ARE IMPORTANT TO INSURE GOOD PERFORMANCE.

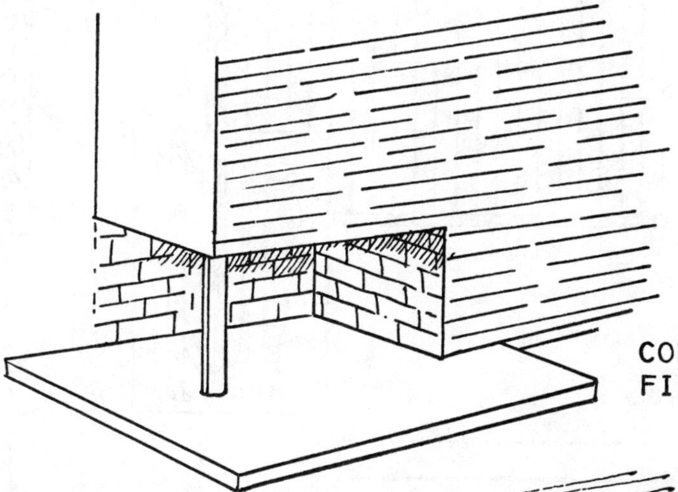

CORNER FIREPLACE

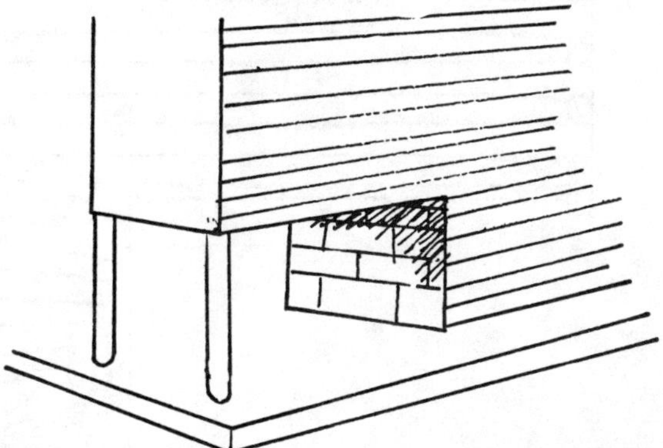

3–SIDED OPENING

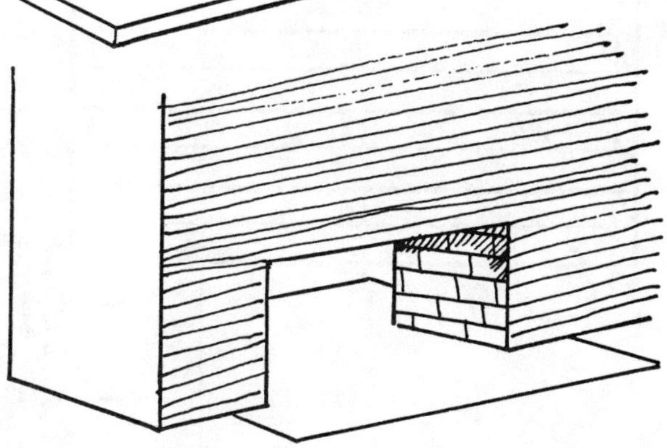

2–SIDED OPENING

FENCE PATTERNS-----

WOOD FENCES ARE USED FOR PRACTICAL PURPOSES AS WELL AS THE ENHANCED BEAUTY OF MANY HOMES. HERE ARE SEVERAL TYPES OF FENCE BUILT FROM STANDARD LUMBER SIZES AND TYPES. TWO TYPES OF PICKET FENCE BUILT OF SPLIT OR HALF ROUND PICKETS ARE ALSO SHOWN.

UNLESS YOU USE REDWOOD, CYPRESS, CEDAR OR OTHER WOOD THAT IS IMPERVIOUS TO WEATHER AND ROTTING, IT IS ADVISABLE TO USE TREATED WOOD TO RESIST DECAY OR DAMAGE BY TERMITES.

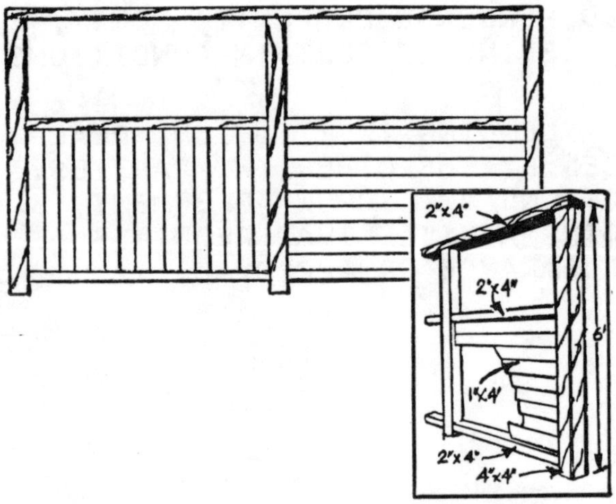

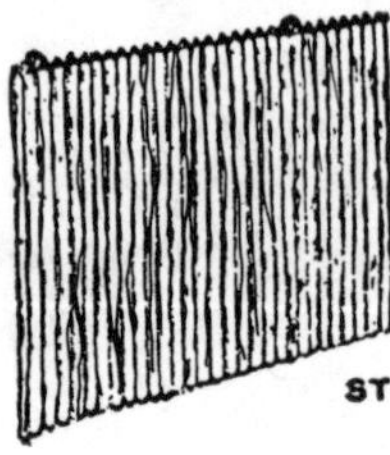
STOCKADE

SPLIT PICKET

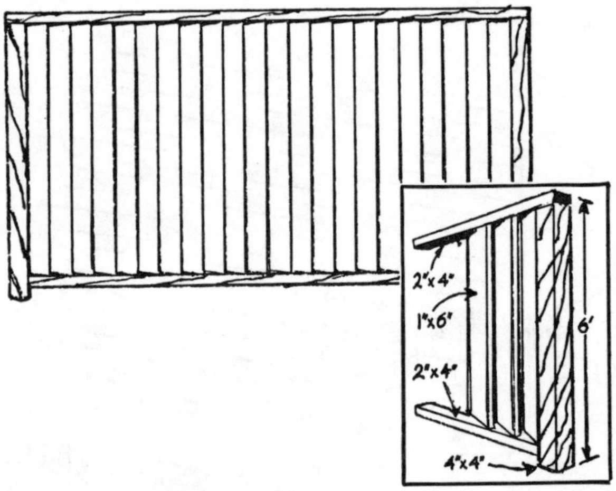

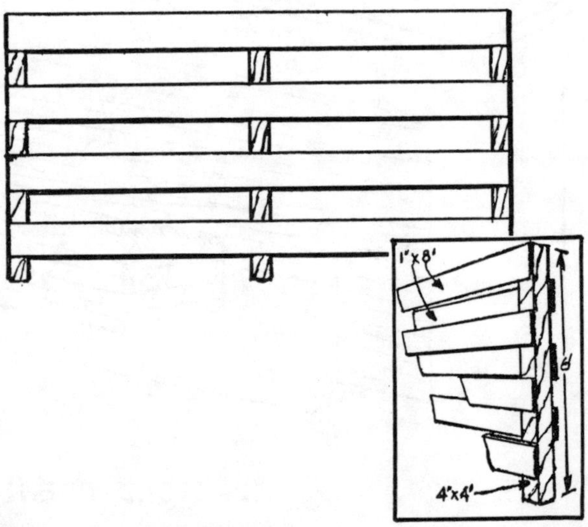

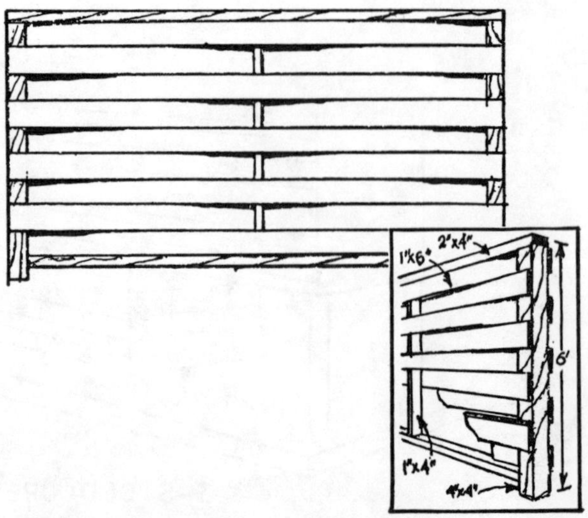

FLOORS-----

A GREAT MANY TYPES OF FLOORING MATERIALS ARE AVAILABLE. THEY INCLUDE SEVERAL KINDS AND TYPES OF WOOD FLOORS FROM PRE-FINISHED OR PLAIN HARDWOODS TO LAMINATED PARQUET OR PLANK TYPES, ALSO PRE-FINISHED AND READY TO USE. FOR HARD SURFACED FLOORS YOU MAY CHOOSE BETWEEN CERAMIC TILE OF VARIOUS TYPES, TERRAZZO, PATIO BLOCK, BRICK OR DECORATED AND COLORED CONCRETE.

A WIDE VARIETY OF COLORS AND PATTERNS ARE AVAILABLE IN VINYL AND ASPHALT FLOOR TILES WHICH CAN BE INSTALLED ON ALMOST ANY TYPE SUBFLOOR SURFACE. VINYL FLOOR COVERINGS ARE ALSO AVAILABLE IN WIDE SHEET ROLLS TO FACILITATE SEAMLESS INSTALLATIONS.

CARPET IS NOW ONE OF THE PREDOMINANT AND UNIVERSALLY USED FLOOR COVERINGS IN TODAY'S HOMES. THE COLORS AND QUALITY, PATTERNS, WEIGHTS, TYPE BACKING, ETC., IN CARPETS IS UNLIMITED.

GOOD QUALITY SUBFLOORS AND/OR UNDERLYING PADDING ARE REQUIRED TO IMPROVE AND LENGTHEN THE LIFE OF CARPETING OR OTHER RESILIENT TYPE FLOORS.

FLOOR TILE COVERAGE-----

THIS CHART WILL ENABLE YOU TO EASILY DETERMINE THE NUMBER OF PIECES REQUIRED FOR VARIOUS FLOOR AREAS.

SELECT NEAREST FLOOR AREA IN COLUMN 1. COLUMNS 2 AND 3 SHOW THE NUMBER OF PIECES OF 9X9 OR 12X12 TILE REQUIRED.

SQUARE FEET OF FLOOR	PIECES 9X9 TILE	PIECES 12X12 TILE
45	80	45
90	160	90
135	240	135
180	320	180
225	400	225
270	480	270
315	560	315
360	640	360
405	720	405
450	800	450
495	880	495
540	960	540
585	1,040	585
630	1,120	630
675	1,200	675
720	1,280	720
765	1,360	765
810	1,440	810
855	1,520	855
900	1,600	900

COURTESY AZROCK FLOOR PRODUCTS DIVISION, MANUAL OF INSTALLATION

FLOOR PLANS-----

FLOOR PLANS OF EACH LEVEL OF A HOUSE ARE A PART OF EACH SET OF HOME PLANS AND ARE REQUIRED TO SECURE BUILDING PERMITS. THEY SHOW THE NAMES AND DIMENSIONS OF ALL AREAS.

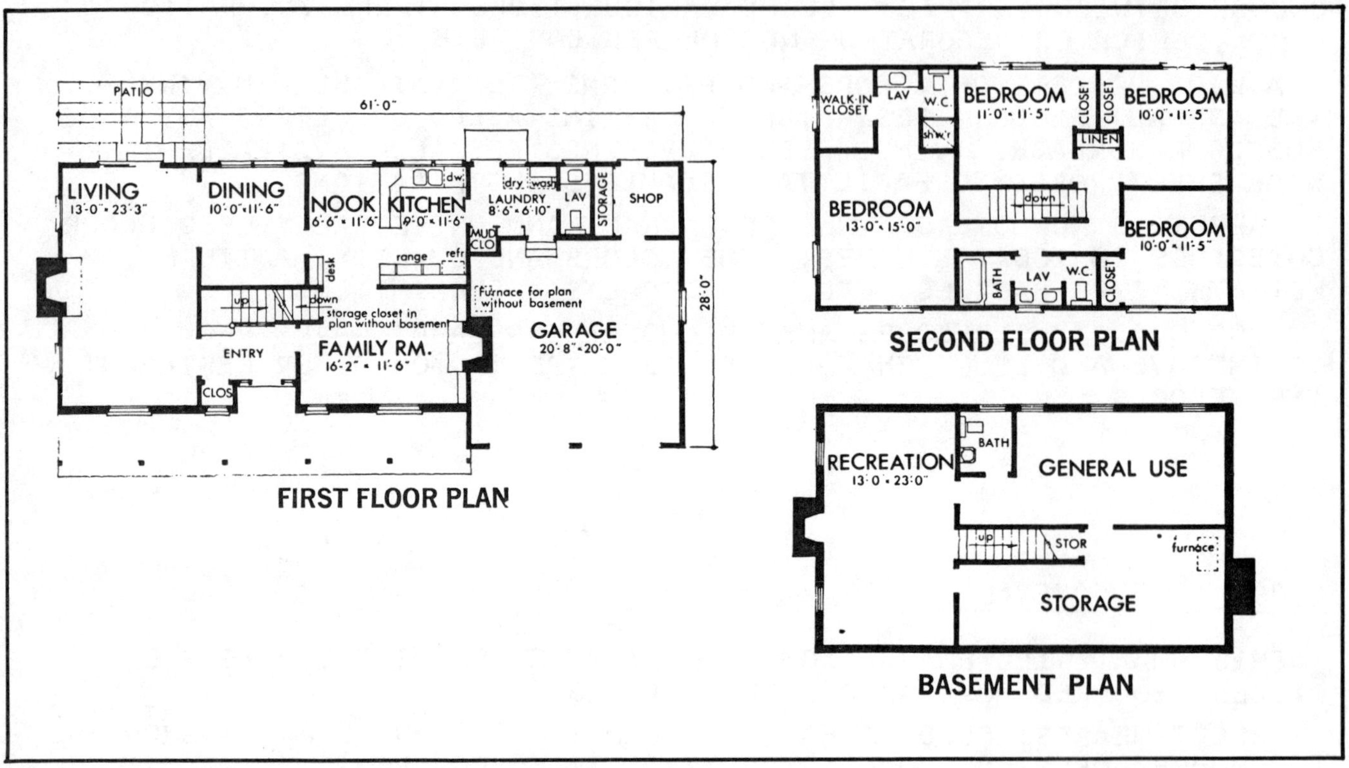

HOUSE PLANS, CROSS SECTION-----

A COMPLETE SET OF HOUSE PLANS INCLUDES A CROSS SECTION AS ILLUSTRATED HERE. THIS TYPE SECTION INDICATES MANY CONSTRUCTION DETAILS NOT SHOWN IN FLOOR PLANS OR ELEVATIONS.

MANY CODES AND BUILDING DEPARTMENTS REQUIRE DETAILED SECTIONS IN ADDITION TO FLOOR PLANS AND ELEVATIONS.

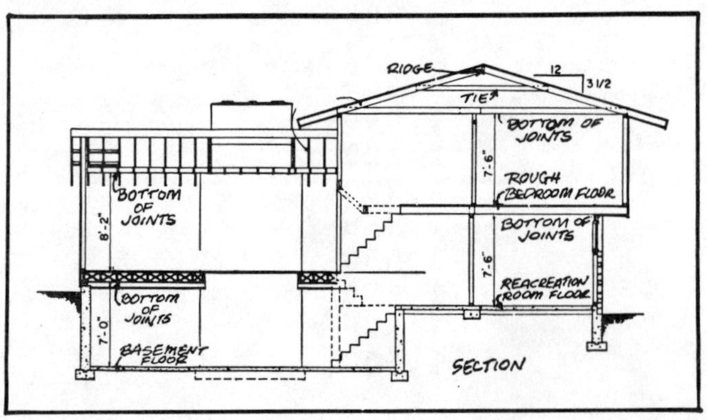

SEE: *ARCHITECTURAL DETAILS.*

GARAGE SIZES-----

WITH THE APPARENT TREND TO SMALLER, MORE COMPACT CARS DUE TO THE ENERGY SITUATION, THE SIZE OF THE GARAGE OR CARPORT MAY VARY FROM PRESENT DAY PLANS.

GARAGE DIMENSIONS SHOWN BELOW SHOULD BE CONSIDERED MINIMAL WITH NO ALLOWANCE FOR TYPICAL CATCH-ALL STORAGE SPACE.

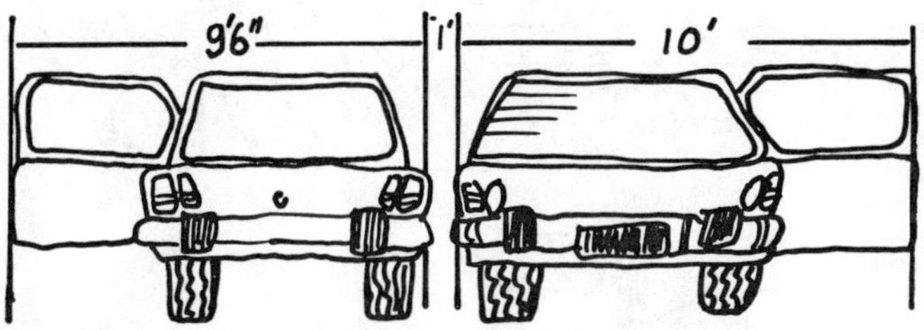

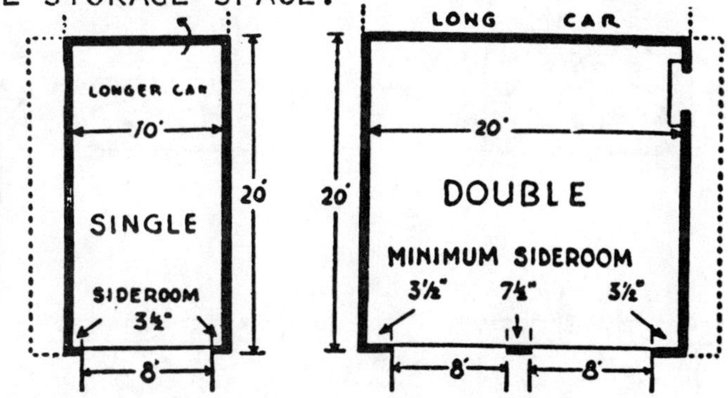

FOUNDATIONS, FOOTERS-----

WHILE FOUNDATIONS AND FOOTERS DO NOT ATTRACT MUCH ATTENTION, THEY ARE THE MOST IMPORTANT PARTS OF A WELL-BUILT HOME. TYPICAL SPREAD FOOTERS OF POURED CONCRETE SUPPORT EITHER CONCRETE OR CONCRETE MASONRY FOUNDATION WALLS IN SKETCHES AT LEFT BELOW. TYPICAL TYPES AND SIZES OF CONCRETE MASONRY UNITS ARE SHOWN.

WALL FOOTER DETAILS

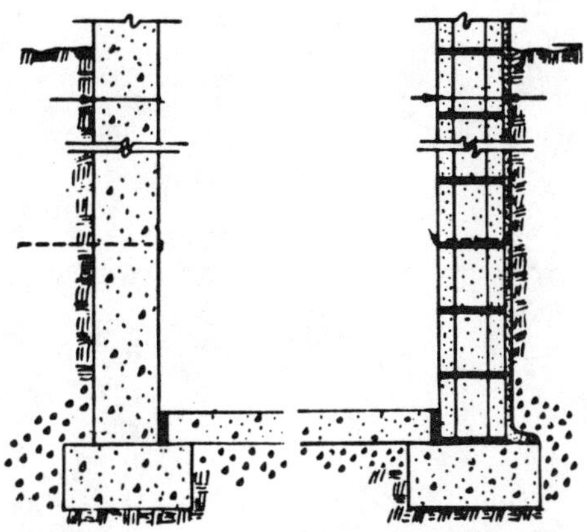

POURED CONCRETE

CONCRETE MASONRY

CONCRETE MASONRY UNITS

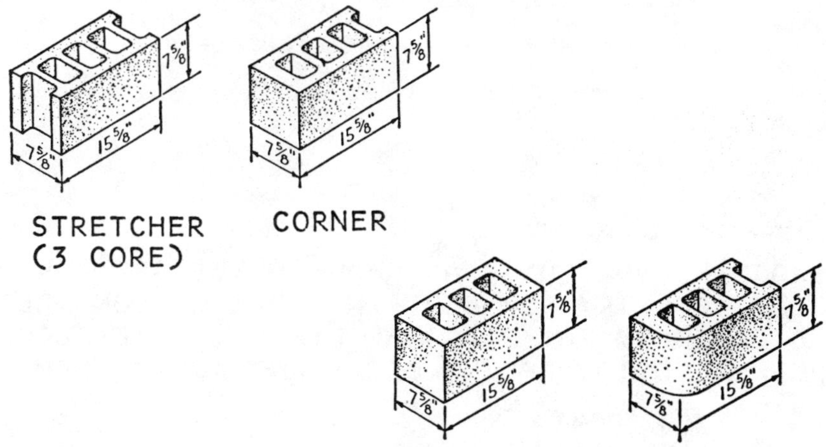

STRETCHER (3 CORE)

CORNER

DOUBLE CORNER OF PIER

BULL NOSE

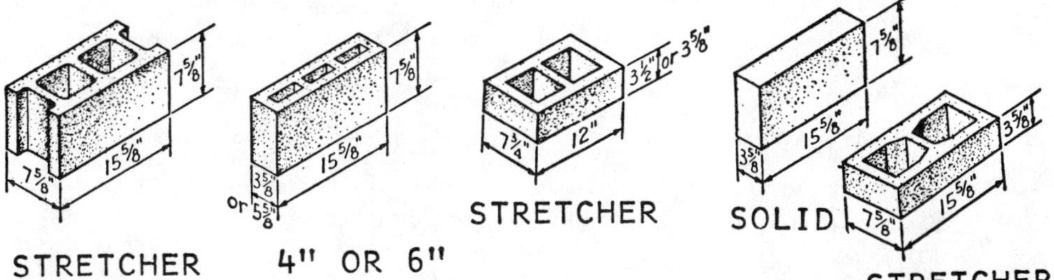

STRETCHER (2 CORE)

4" OR 6" PARTITION

STRETCHER

SOLID

STRETCHER (MODULAR)

COURTESY NATIONAL CONCRETE MASONRY ASSOC.

FOOTER FOUNDATION, DETAILS-----

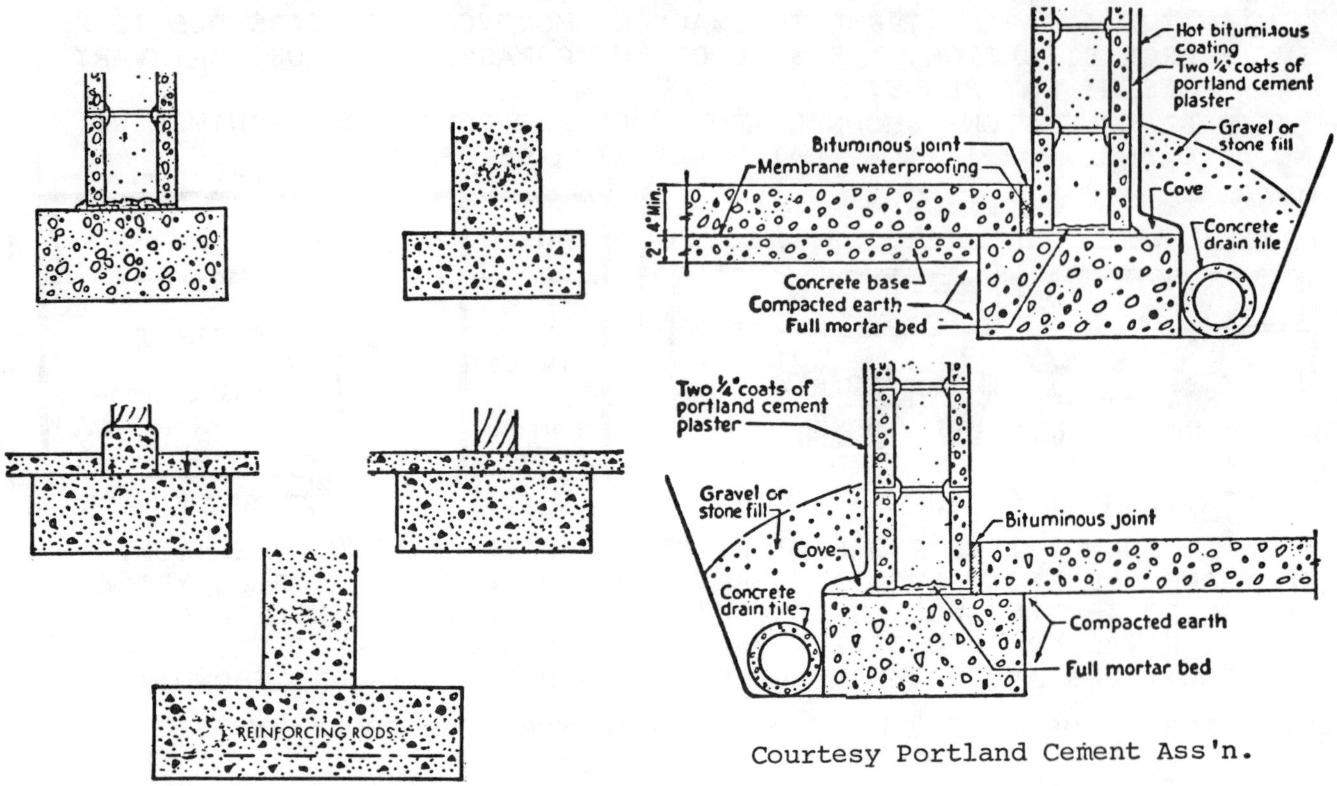

Courtesy Portland Cement Ass'n.

FOUNDATIONS, WOOD-----

WOOD FOUNDATIONS, AS SHOWN HERE, ARE NOW APPROVED IN SOME AREAS OF THE COUNTRY FOR CERTAIN TYPES OF CONSTRUCTION. THE WOOD PARTS INDICATED BY THE DIAGONAL LINES IN THE ACCOMPANYING SKETCHES ARE PRETREATED TO PREVENT DECAY AND AVOID TERMITES.

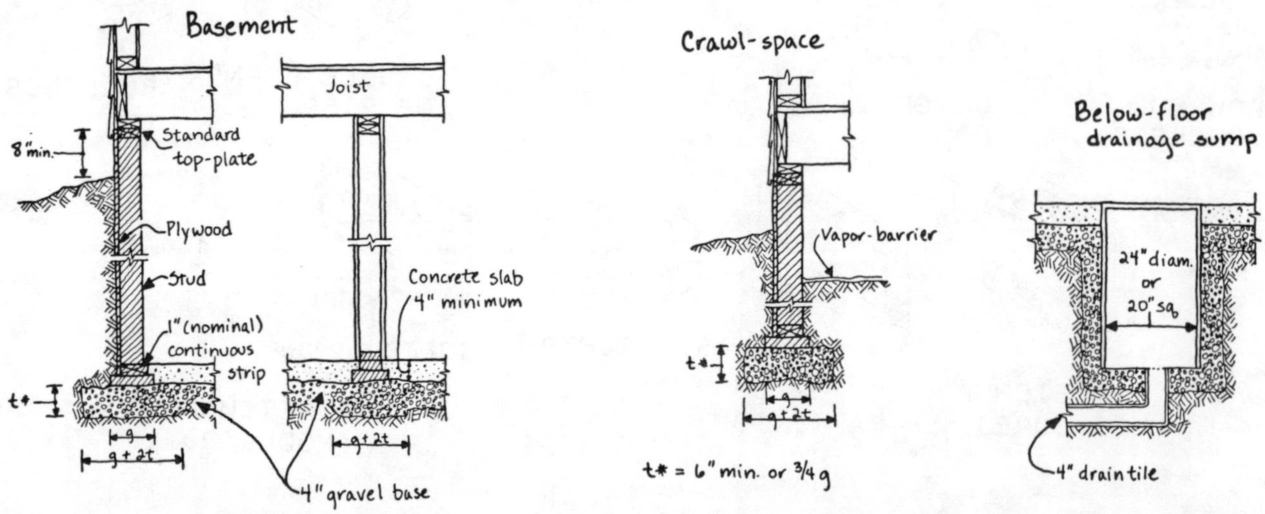

COURTESY SUCCESSFUL HOW TO BUILD YOUR OWN HOME, BY ROBERT C. RESCHKE (STRUCTURES PUBLISHING CO.)

HOUSE PLANS-----

CONSTRUCTION PLANS, SPECIFICATIONS AND MATERIAL LISTS, AS SHOWN HERE, ARE FURNISHED WHEN YOU BUILD HOUSE PLANS FROM ONE OF THE SEVERAL RELIABLE DESIGN CONCERNS IN THIS FIELD.

BLUEPRINTS, LISTS OF MATERIALS, AND SPECIFICATIONS ARE SUPPLIED TO SIMPLIFY THE MATTER OF SECURING BIDS ON MATERIALS AND LABOR.

THE FOLLOWING FIRMS ARE RECOGNIZED LEADERS IN THIS FIELD:

HOME BUILDING PLAN SERVICE
PORTLAND, OREGON

MASTER PLAN SERVICES
MINEOLA, NEW YORK

HOME PLANNERS, INC.
FARMINGTON, MICHIGAN

GARLINGHOUSE HOME PLANS
TOPEKA, KANSAS

CONSTRUCTION PLANS

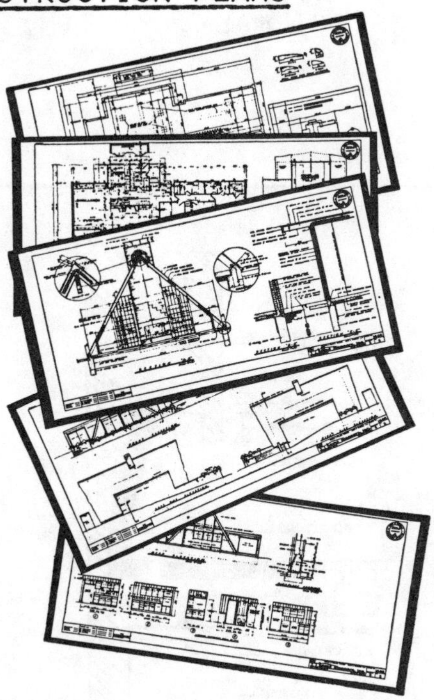

MATERIALS LIST

MASONRY, VENEER & FIREPLACE - FRAMING LUMBER - ROOFING AND SHEET METAL - WINDOWS AND DOOR FRAMES - EXTERIOR TRIM AND INSULATION - TILE WORK, FINISH FLOORS - INTERIOR TRIM, KITCHEN CABINETS - ROUGH AND FINISH HARDWARE.

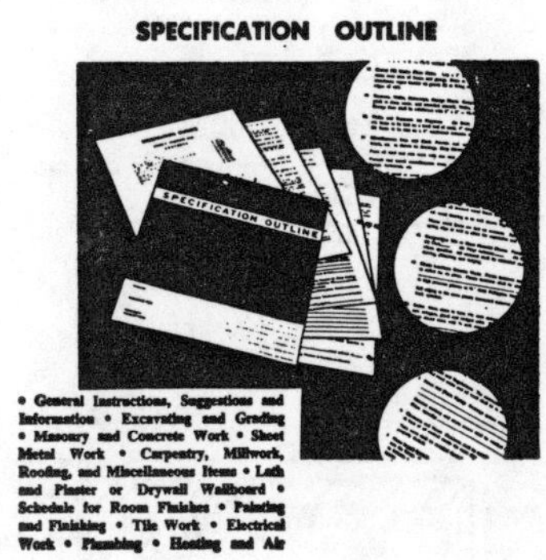

SPECIFICATION OUTLINE

GENERAL INSTRUCTIONS, SUGGESTIONS AND INFORMATION - EXCAVATING AND GRADING - MASONRY AND CONCRETE WORK - SHEET METAL WORK - CARPENTRY, MILLWORK, ROOFING, AND MISCELLANEOUS ITEMS - LATH AND PLASTER OR DRYWALL WALLBOARD - SCHEDULE FOR ROOM FINISHES - PAINTING AND FINISHING - TILE WORK - ELECTRICAL WORK - PLUMBING - HEATING AND AIR CONDITIONING.

FIREPLACES-----

THE FIREPLACE IS ONE OF THE MOST POPULAR PARTS OF A HOME. IT MAY BE INCLUDED AS THE HOME IS BUILT, ADDED LATER OR INCLUDED IN MODERNIZATION. THESE SKETCHES SHOW THE BASIC PARTS OF A FIREPLACE.

PREFABRICATED METAL FIREPLACE UNITS ARE AVAILABLE IN CIRCULATING AND NON-CIRCULATING TYPES. THEY ARE USUALLY USED IN CONJUNCTION WITH MASONRY FACING, BACKWALL AND CHIMNEY CONSTRUCTION. SOME TYPES OF METAL FIREPLACE UNITS, HOWEVER, ARE DESIGNED FOR ZERO CLEARANCE FOR ADJACENT WOOD CONSTRUCTION AND MAY BE INSTALLED IN WOOD FRAMED CHASES IN WHICH FIREPLACE VENTING IS PROVIDED BY SUITABLE METAL CHIMNEYS. CIRCULATING UNITS USE NATURAL CIRCULATION OR ELECTRIC FANS TO SUPPLY MORE HEAT TO ADJACENT ROOMS.

NEWER TYPES OF ENERGY CONSERVING METAL FIREPLACE UNITS NOW ARE AVAILABLE. THEY CIRCULATE ROOM AIR BUT BRING IN OUTDOOR AIR FOR COMBUSION PURPOSES, KEEPING THE TWO AIR CIRCULATION PATHS SEPARATED BY THE USE OF GLASS DOORS ON THE FIREPLACE OPENING.

MOST FIREPLACES TODAY ARE INTENDED FOR WOOD BURNING USE. THE PROVISION OF GAS LOG LIGHTERS OR GAS/ELECTRIC LOGS HAS DIMINISHED GREATLY WITH RISING UTILITY RATES.

THE FOLLOWING COMPANIES ARE THE LEADING MANUFACTURERS OF PREFABRICATED FIREPLACE UNITS: HEATILATOR, MAJESTIC, MALM, MARTIN, PREWAY, SUPERIOR AND TEMCO.

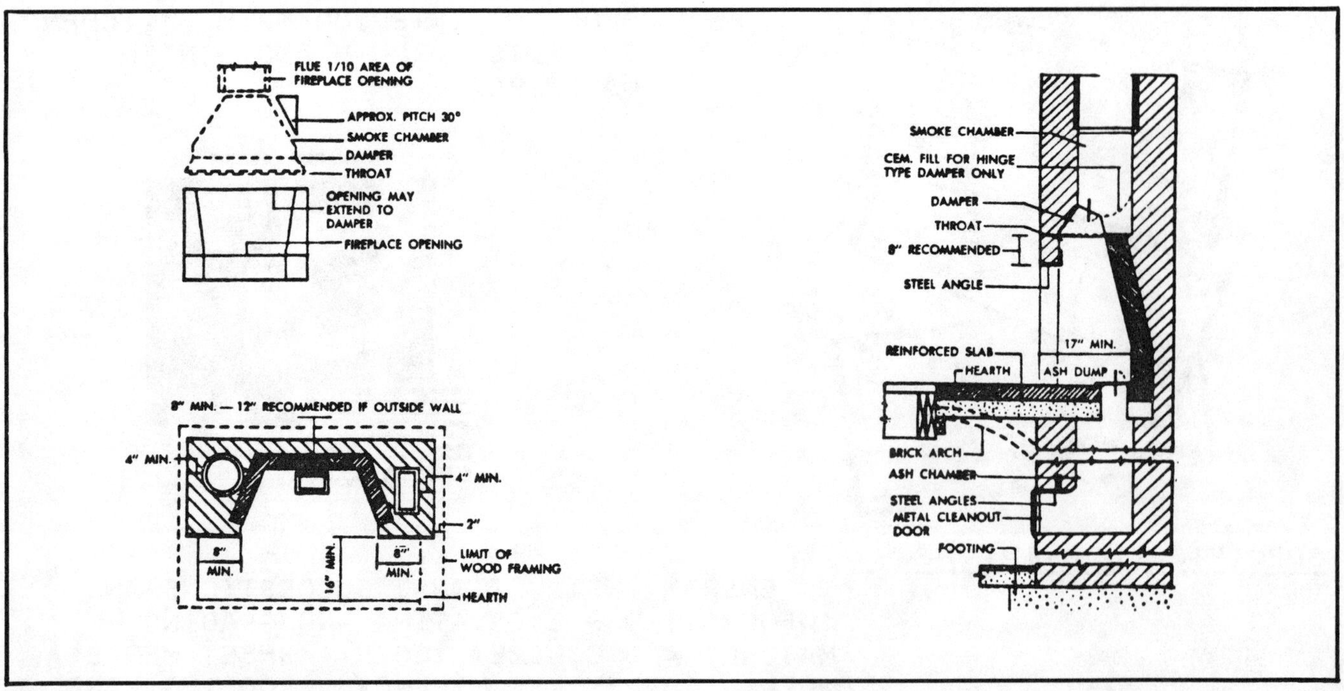

COURTESY <u>BOOK OF SUCCESSFUL FIREPLACES</u> (STRUCTURES PUBLISHING CO.)

FREE STANDING FIREPLACES-----

MODERN FREE STANDING FIREPLACES WITH PREFABRICATED METAL CHIMNEYS ARE NOW AVAILABLE AND PERMISSIBLE IN MOST NEW AND PRESENT HOMES.

THE SKETCH AT THE RIGHT SHOWS THE INSTALLATION OF THIS TYPE UNIT IN A HOUSE WITH A NORMAL TYPE CEILING AND PITCHED ROOF.

MANY TYPES AND STYLES ARE AVAILABLE TO SUIT MOST CONDITIONS AND MAY BE ADAPTED TO VARIOUS STRUCTURAL ARRANGEMENTS AND ROOF DESIGNS.

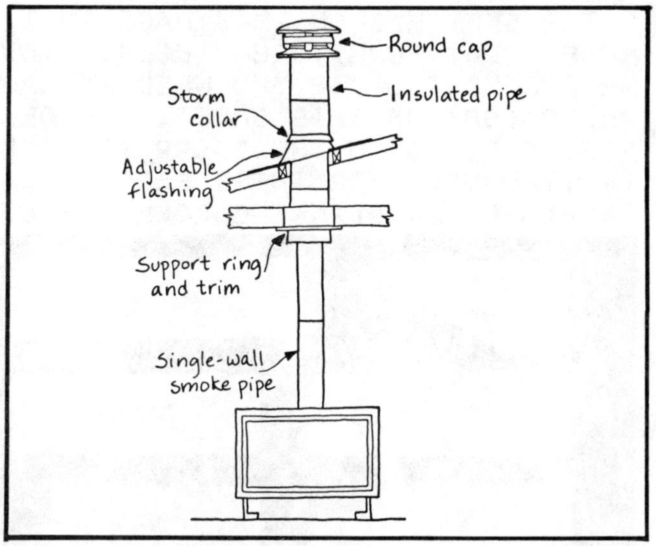

COURTESY <u>SUCCESSFUL HOW TO BUILD YOUR OWN HOME</u>, BY ROBERT C. RESCHKE (STRUCTURES PUBLISHING CO.)

GLASS-----

WITH RESPECT TO ENERGY LOSS THROUGH GLASS, A TYPICAL 3' BY 5' WINDOW, SINGLE GLAZED AND DOUBLE HUNG, LOSES ABOUT THREE TIMES AS MUCH ENERGY BY CONDUCTION THROUGH THE GLASS AS IT DOES VIA LEAKS AROUND THE PERIMETER EVEN WHEN THE LATTER HAS HAD FRAME CRACKS SEALED OR INSULATED DURING INSTALLATION.

REDUCTION OF WINDOW AREA IS OBVIOUSLY AN EASY WAY TO REDUCE ENERGY LOSSES THROUGH GLASS. A COMPARISON WITH A REASONABLY INSULATED FRAME WALL INDICATES THAT HEAT LOSSES THROUGH SINGLE PANE GLASS MAY BE TEN TIMES GREATER THAN THROUGH A WALL AREA OF THE SAME SIZE.

HEAT LOST OR GAINED THROUGH GLASS DEPENDS UPON THE DIFFERENCE BETWEEN INDOOR AND OUTDOOR AIR TEMPERATURES. THIS DIFFERENCE IS USUALLY GREATER DURING WINTER IN COLD REGIONS THAN IT IS IN SUMMER IN WARM REGIONS. THUS, INSULATING GLASS TENDS TO BE OF GREATER IMPORTANCE DURING WINTER IN COLDER CLIMATES ALTHOUGH IT IS STILL VERY HELPFUL IN REDUCING INTERIOR HEAT GAINS IN WARM WEATHER AREAS.

TINTED GLASS, BECAUSE IT REDUCES BOTH CONDUCTIVE TRANSMISSION THROUGH THE GLASS AS WELL AS SOLAR RADIATION TO THE INTERIOR, CAN BE EFFECTIVE AS AN AIR CONDITIONING ENERGY SAVER IN WARM CLIMATES.

THE USE OF STORM WINDOWS HAS NEARLY THE SAME REDUCTION IN HEAT LOSSES THROUGH WINDOWS AS THE ADDING OF A SECOND PANE OF GLASS. HEAT TRANSMISSION AND THERMAL RESISTANCE VALUES FOR COMMON TYPES OF GLAZING MATERIALS ARE SHOWN BELOW.

TYPE OF GLAZING	GLASS THICKNESS	AIR SPACE	U VALUE	R VALUE
SINGLE PANE, CLEAR	1/4" OR 1/8"	---	1.13	0.88
SINGLE PANE, TINTED	"	---	1.00	1.00
DUAL PANE, CLEAR	"	1/2"	0.6	1.66

FURNITURE SKETCHES-----

THE SKETCHES ON THIS PAGE IN 1/4" SCALE SIZE COVER MOST OF THE MODERN DAY FURNITURE PIECES. YOU CAN USE THESE TO MAKE ROOM ARRANGEMENTS. IF YOU NEED EXTRA PIECES YOU CAN EASILY MAKE A PHOTO COPY OF THIS SHEET. IF DESIRED, YOU MAY COLOR THE VARIOUS PIECES TO MATCH THE DECOR YOU EXPECT TO USE. YOU WILL FIND UNLIMITED USES FOR THESE SKETCHES AS WELL AS THE GRAPH SHEETS AND TEMPLATES IN PLANNING ROOM LAYOUTS AND ARRANGEMENTS.

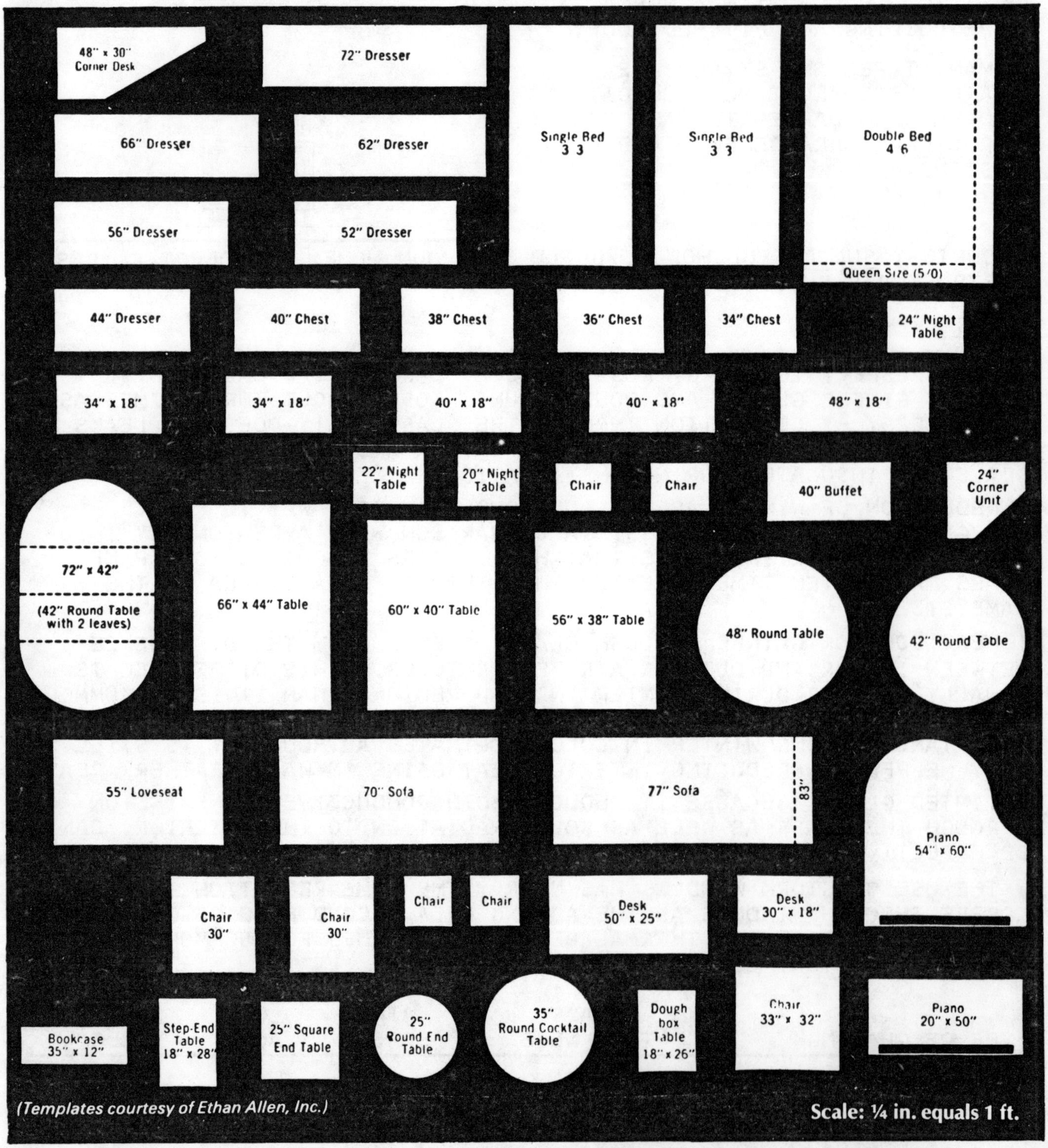

SEE: *TEMPLATES, GRAPH SHEET, SKETCHING, ROOM ARRANGEMENTS.*

FRAMING-----

MANY PARTS AND TERMS ASSOCIATED WITH HOUSE FRAMING ARE DESCRIBED IN THE ACCOMPANYING DIAGRAM.

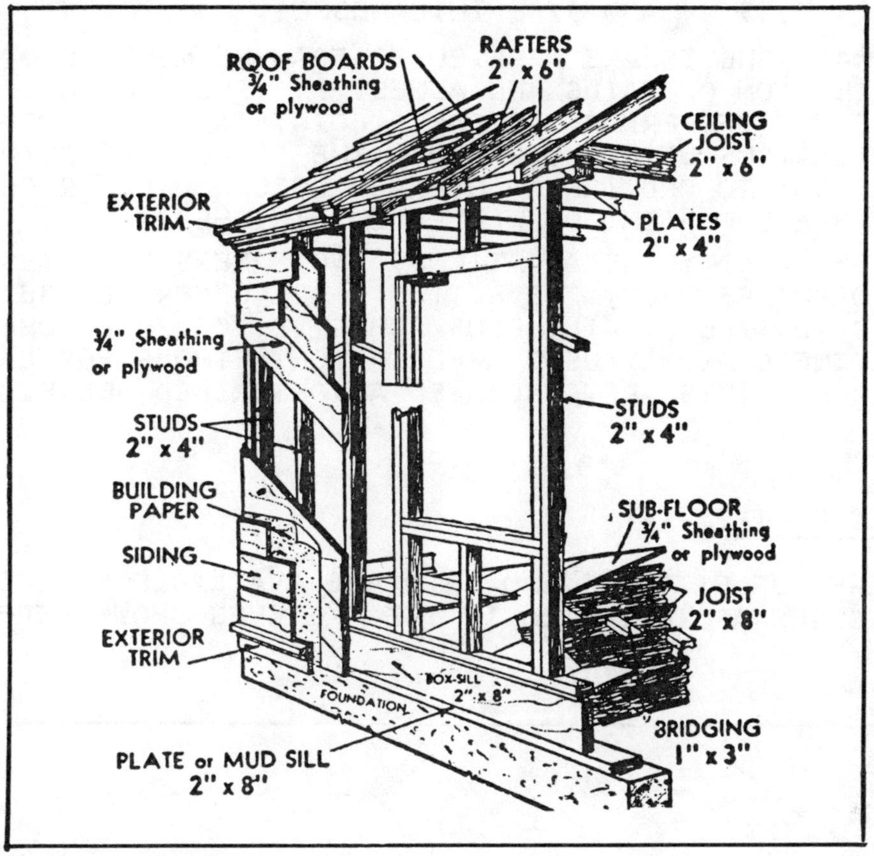

RESTRICTIONS ON BUILDING-----

RESTRICTIONS COVERING ANY AND ALL REGULATIONS ARE USUALLY RECORDED AND SHOWN ON ALL DEEDS FOR PROPERTY. FURTHER DATA AND INFORMATION ON MATTERS OF THIS TYPE ARE AVAILABLE FROM THE OFFICES OR AUTHORITIES WHO HAVE JURISDICTION OVER BUILDING CODES AND PERMITS IN THE AREA IN QUESTION. QUITE OFTEN RESTRICTIONS ARE OUTMODED AND REQUIRE AN APPEAL OR CHANGE TO MEET NEW CONDITIONS. IN CASES OF THIS TYPE NO ACTION SHOULD BE TAKEN UNTIL PROPER AUTHORITY IS GIVEN AND FULLY APPROVED.

Notes: ______________________________

GYPSUM DRYWALL-----

GYPSUM WALLBOARD IS USED FOR THE SURFACE LAYER ON INTERIOR WALLS AND CEILINGS. IT IS MADE IN 4' WIDTHS AND IN LENGTHS OF 8', 10', 12' AND 14'. IT IS COMMONLY USED IN 1/2" THICKNESS BUT IS ALSO PRODUCED IN 1/4", 3/8" AND 5/8" THICKNESSES.

GYPSUM DRYWALL SHEETS ARE APPLIED TO FRAMING MEMBERS BY NAILING OR BY A COMBINATION OF NAILS AND ADHESIVE. SHEET JOINTS ARE CONCEALED BY THE USE OF PERFORATED PAPER JOINT TAPE EMBEDDED IN TWO OR MORE APPLICATIONS OF JOINT CEMENT, THE FINAL COAT BEING FEATHER EDGED OUT TO PROVIDE AN OVERALL SMOOTH WALL OR CEILING SURFACE SUITABLE FOR PAINT OR WALLPAPER FINISHES.

IN SOME HIGH QUALITY INSTALLATIONS, TWO LAYERS OR PLIES OF GYPSUM WALLBOARD ARE USED, THE BOTTOM LAYER GENERALLY BEING OF THE BACKER BOARD TYPE. NAILS PLUS ADHESIVE, OR FULL ADHESIVE COATING, ARE THE COMMONLY USED APPLICATION METHODS FOR LAYERING. THE TOP LAYER'S JOINTS ARE CONCEALED AS DESCRIBED ABOVE.

GYPSUM WALLBOARD TRIM-----

VARIOUS TYPES OF FITTINGS AND TRIM ARE AVAILABLE FOR JOINTS, CORNERS AND EDGES AS INDICATED IN THE SKETCHES SHOWN HERE.

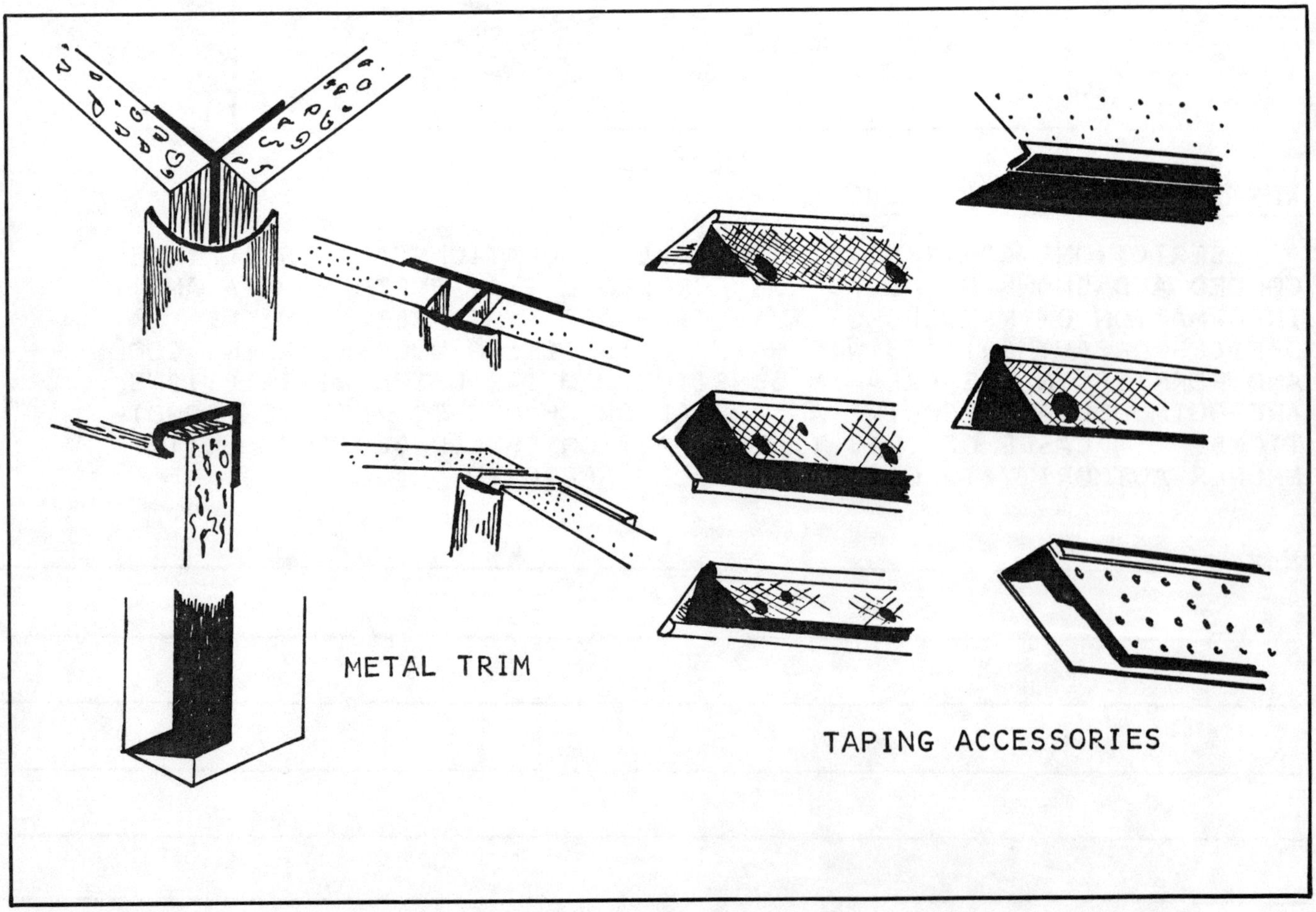

GYPSUM, DRYWALL TYPES-----

COMMON FACTORY PRODUCED TYPES OF GYPSUM WALLBOARD ARE LISTED BELOW BUT NOT ALL SUPPLIERS OF GYPSUM WALLBOARD PRODUCTS CARRY ALL TYPES IN STOCK.

REGULAR: HAS GRAY LINER PAPER ON THE BACK. THE FACING SIDE HAS A SPECIAL CREAM COLORED PAPER THAT PROVIDES A SMOOTH AND EVEN SURFACE FOR DECORATION.

INSULATING: HAS BRIGHT FINISHED ALUMINUM FOIL BONDED TO THE GRAY LINER PAPER ON THE BACK.

PREDECORATED: HAS A DECORATIVE VINYL OR PAPER SHEET BONDED TO THE EXPOSED FACE SURFACE WHICH DOES NOT REQUIRE FURTHER DECORATIVE TREATMENT.

TYPE X: HAS MORE FIRE RESISTANT MATERIALS IN CORE FOR USE IN MANY FIRE RESISTANT RATED CONSTRUCTION ASSEMBLIES.

BACKING BOARD: HAS GRAY LINER PAPER ON BOTH SIDES AND IS USED IN MULTI-PLY CONSTRUCTION FOR EXTRA FIRE RESISTANCE, SOUND ISOLATION OR WALL STRENGTH. BACKING BOARD IS OBTAINABLE IN INSULATING, TYPE X FIRE RESISTANT OR WATER RESISTANT TYPES. THE LATTER IS SUITABLE AS A BASE FOR WALL TILE IN BATHS, SHOWERS OR OTHER WET AREAS.

GYPSUM LATH-----

GYPSUM LATH, AS USED FOR PLASTER BASE, IS MADE IN FOUR TYPES. THIS MATERIAL IS UNIVERSALLY USED WHEN PLASTER IS REQUIRED.

GYPSUM PLASTER LATH IS NAILED DIRECTLY TO THE STUDS OR JOISTS. SIZES USUALLY 16" X 48" ACCOMMODATE 12" OR 16" CENTERS.

PLAIN: GYPSUM LATH PLASTER BASE IN SHEETS. SURFACED WITH A FIBROUS PAPER SUITABLE FOR PLASTER APPLICATION.

PERFORATED: PERFORATED LATH IS REGULAR TYPE WITH 3/4" DIAMETER HOLES 4" ON CENTER.

INSULATING: INSULATING LATH IS LATH WITH ALUMINUM FOIL BONDED TO THE BACK OR UNEXPOSED FACE, PROVIDING A VAPOR BARRIER AND A REFLECTIVE INSULATOR.

TYPE X: SAME AS ABOVE WITH FIRE RESISTANT CORE MATERIALS.

Notes: ______________________________

GYPSUM WALLBOARD-----

VARIOUS EDGE TYPES ARE AVAILABLE IN MOST AREAS. THE TYPE YOU WILL REQUIRE WILL DEPEND ON THE JOINT TREATMENT YOU SELECT.

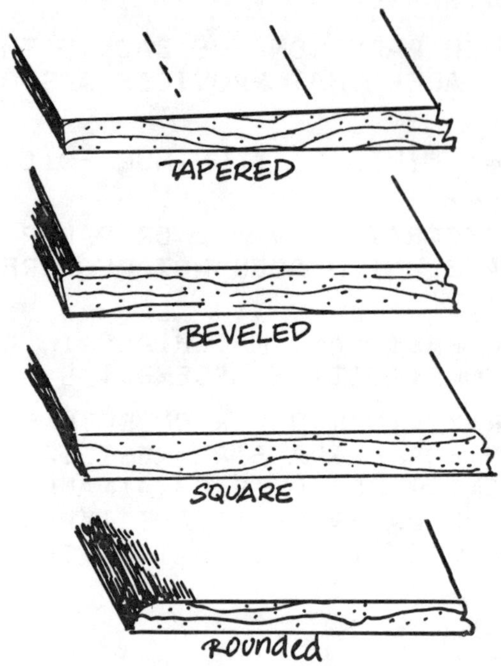

COURTESY USING GYPSUMBOARD FOR WALLS AND CEILINGS BY THE GYPSUM ASSOCIATION

GARAGE PARTS-----

GARAGES AND CARPORTS ARE IMPORTANT TO EVERY HOME FOR THE PROTECTION OF AUTOS. THIS SKETCH SHOWS YOU THE VARIOUS PARTS OF A TYPICAL GARAGE.

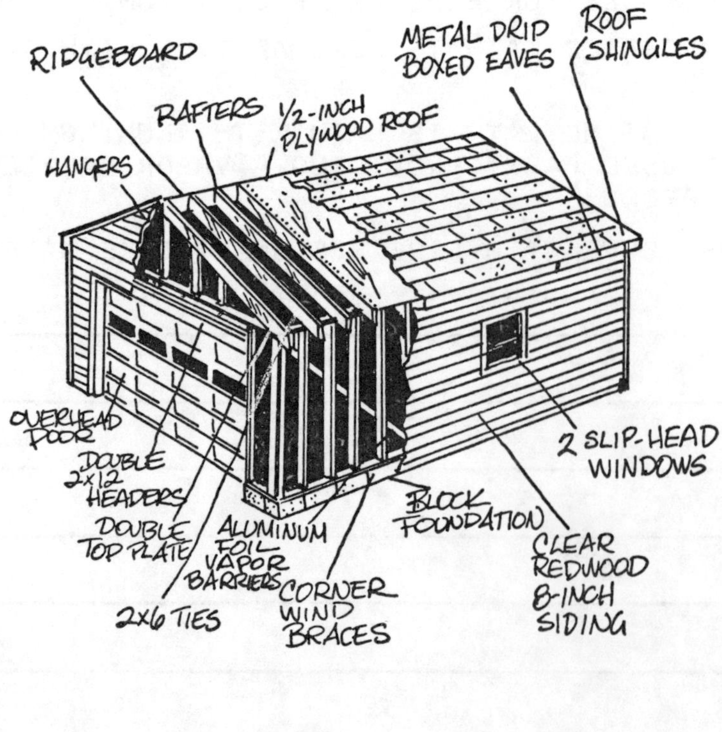

GARAGE COMBINATIONS-----

MANY MODERN DAY GARAGES AND CARPORTS SERVE SEVERAL USEFUL PURPOSES. STORAGE AREAS, WORK AREAS, SPACE FOR BOATS, BIKES, ETC., ARE OFTEN INCLUDED IN THE PLANS. PERHAPS YOU CAN FIND A FLOOR PLAN HERE THAT WILL BEST SERVE YOUR NEEDS. ADJUST THE DIMENSIONS TO PROVIDE SPACE FOR YOUR EQUIPMENT.

HARDBOARD-----

ONE OF THE MOST VERSATILE PRODUCTS FOR USE IN HOME CONSTRUCTION AND REMODELING IS COMMONLY KNOWN AS HARDBOARD. IT IS MADE IN SMOOTH PANELS BOTH PLAIN AND DECORATED IN MANY PATTERNS AND STYLES. IT IS NORMALLY 1/8" OR 1/4" THICK IN 4' X 8' SIZED PANELS. HARDBOARD PANELS ARE AVAILABLE IN TEMPERED AND UNTEMPERED TYPES WITH VARIOUS SURFACE TEXTURES. IT IS ALSO MADE IN PERFORATED OR PEG BOARD TYPES AND IN VARIOUS SIDING PANELS FOR EXTERIOR APPLICATIONS.

SEE: *HARDBOARD MOLDINGS.*

Notes: ______________________________

GUTTER PARTS-----

GUTTERS ARE GENERALLY REQUIRED ALTHOUGH IN SOME AREAS ROOF WATER IS PERMITTED TO RUN OFF ONTO THE GROUND. GUTTERS MAY BE MADE OF WOOD, COPPER, ALUMINUM, GALVANIZED OR PAINTED STEEL, OR VINYL. THEY COME READY PAINTED, COATED OR IN NATURAL WOOD. THE MANY PARTS OF THE FRONT EDGE SHAPE (O.G.) ARE SHOWN HERE.

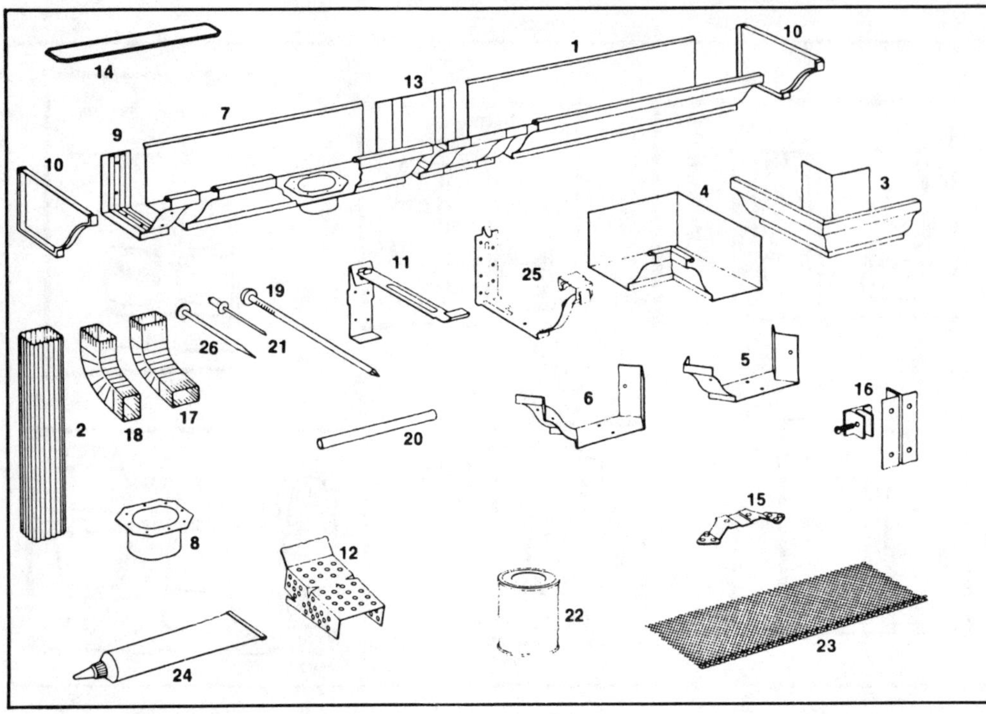

KEY

1 5" OG Gutter
2 2"x3" Square Downspout
3 Miter Section Outside
4 Miter Section Inside
5 Miter Strip Outside
6 Miter Strip Inside
7 Section with Outlet Tube
8 Wide Flange Outlet Tube
9 Slip Joint Connector
10 End Caps-Tap On
11 Fascia Bracket
12 Perforated Strainer
13 Expansion Joint
14 Downspout Pipe Band
15 Downspout Pipe Cleat
16 Downspout Pipe Bracket
17 2"x3" Square Elbow Style A 60° - 75°
18 2" x 3" Square Elbow Style B 75°
19 Spike
20 Ferrule
21 Rivets
22 Touch Up Paint
23 Gutter Cover
24 Joint Sealer
25 Fascia Bracket w/Spring Clip
26 Aluminum Nail

COURTESY KAISER ALUMINUM

HOMEOWNERS WARRANTY-----

IN MANY SECTIONS OF THE COUNTRY SOME MEMBERS OF THE NATIONAL ASSOCIATION OF HOMEBUILDERS PROVIDE A HOME OWNERS WARRANTY (H.O.W.) ON NEW HOMES THEY BUILD OR SELL. THIS WARRANTY, WHEN AVAILABLE, PROTECTS YOU UNDER CERTAIN CONDITIONS FROM 1 TO 10 YEARS AGAINST FAILURE ON THE PART OF THE BUILDER, MATERIALS AND WORKMANSHIP.

Notes:

GRAPH SHEET (1/8")-----

THIS SHEET MAY BE USED FOR SKETCHING PURPOSES BUT IT IS PRIMARILY FOR USE WHERE REAL PROPERTY OR TRACTS ARE CONCERNED. USUALLY AN ENGINEER'S SCALE OF 50, 100 OR 200 FEET TO THE INCH IS INDICATED.

GRAPH SHEET (1/4")-----

USE THIS SHEET FOR VARIOUS TYPES OF HOME PLANNING AND SKETCHING. THE 1/4" RULING COORDINATES WITH THE TEMPLATES, FURNITURE CUT OUTS AND SKETCHES IN THIS BOOK. ADDITIONAL COPIES MAY BE MADE BY THE PHOTO COPY PROCESS.

HOUSE COMPONENTS, MATERIALS, ESTIMATING-----

THIS LIST OF COMPONENT PARTS OF A HOUSE WILL HELP YOU UNDERSTAND WHERE AND HOW DIFFERENT MATERIALS ARE USED AND HOW THEY ARE ESTIMATED.*

PART	MATERIALS OR PRODUCTS REQUIRED	COMMON USE OF MEASURE
CEILINGS	GYPSUM BOARD, GYPSUM LATH, INSULATING LATH, ACOUSTICAL TILE, METAL LATH, PLASTER, PLYWOOD, WALLBOARD, PAINT, TRANSLUCENT AND SUSPENDED CEILING	PER 100 FEET (PLASTER PER 100 SQUARE YARDS)
CHIMNEY METAL CHIMNEY UNITS	BRICK, CHIMNEY BLOCK, CONCRETE MASONRY, FLUES, DAMPERS, CLEAN OUT DOORS,METAL	PER FOOT IN HEIGHT
DRAINAGE, SEWERS DITCHING FOOTERS	EXCAVATION, DITCHES, CULVERTS, DRAIN TILE, SEWER PIPE, FORMS, CONCRETE, REINFORCING	PER 100 LINEAR FEET
FIREPLACE	COMMON BRICK, FACE BRICK, FIRE BRICK, DAMPERS, FLUES, GRATE, MANTEL, FACING TILE, METAL FIREPLACE UNITS, SCREEN	PER UNIT
FLOORING	CONCRETE, TILE, BRICK, FLOORING OR FLOOR COVERINGS, CARPET, PLYWOOD, SHEATHING, UNDERLAYMENT, SLATE, TERRAZZO	PER 100 SQUARE FEET
FOUNDATIONS	CONCRETE MASONRY, CONCRETE, STRUCTURAL TILE, FORMS, EXCAVATION, WATERPROOFING, DRAIN TILE, WOOD FOUNDATIONS	PER 100 SQUARE FEET, LINEAR FEET OR CUBIC YARDS
FRAMING	BEAMS, COLUMNS, JOISTS, STUDS, RAFTERS, ROOF AND FLOOR TRUSSES	PER 100 SQUARE FEET
INSULATION	BATTS, BLANKETS, LOOSE FILL TYPES, PLASTIC FOAM, FOAMBOARD, REFLECTIVE	PER 100 SQUARE FEET
PARTITIONS	GYPSUM BOARD (DRYWALL), STUDS, GYPSUM LATH, PLASTER, INSULATING BOARD, GYPSUM TILE, WALLBOARD, PLYWOOD, METAL LATH	PER 100 SQUARE FEET
PLASTERING	GYPSUM LATH, METAL LATH, BEADS, SCREED ACOUSTICAL PLASTER, INSULATING LATH	PER 100 SQUARE YARDS

(CONTINUED)

PART	MATERIALS OR PRODUCTS REQUIRED	COMMON USE OF MEASURE
ROOFS	ASBESTOS, ASPHALT, BUILT-UP, METAL, TILE, SLATE, WOOD SHINGLE ROOFING MATERIALS, EAVES, FELTS, SOFFIT, FLASHINGS, GUTTERS, CONDUCTORS	PER 100 SQUARE FEET "SQUARE"
WALLS	BRICK, CONCRETE, CONCRETE MASONRY, GLASS BLOCK, INSULATION, GYPSUM BLOCK, SHEATHING, PLYWOOD, WALL-BOARD, INSULATION BOARD, PLANK, SIDING, STUDS, STUCCO	PER 100 SQUARE FEET. BOARD FEET, (STUCCO PER 100 SQUARE YARDS)

* MATERIALS ARE FIGURED AND QUOTED ON AN INDIVIDUAL BASIS ALTHOUGH THEY ARE AN INTEGRAL PART OF THE COMPONENT PART. FOR EXAMPLE, A CHIMNEY MAY BE FIGURED PER FOOT IN HEIGHT ON AN OVER-ALL BASIS. BUT IT WILL CONTAIN A CERTAIN NUMBER OF BRICK WHICH ARE FIGURED ON A PER 1,000 BASIS AND SO MANY FEET OF FLUE LINER ON A PER PIECE OR PER FOOT BASIS. 1,000 BRICK WOULD LIKEWISE RE-QUIRE A CERTAIN AMOUNT OF MORTAR FOR LAYING WHICH IS FIGURED ON THE NUMBER OF BAGS OF MORTAR AND AMOUNT OF SAND REQUIRED.

USUALLY HEATING, PLUMBING AND WIRING ARE QUOTED AND SOLD ON A PER JOB BASIS. ITEMS LIKE DOORS, WINDOWS, KITCHEN UNITS, ETC. ARE FIGURED AND QUOTED ON A PER UNIT BASIS.

SEE: *HOUSE PARTS*.

INSURANCE-----

INSURANCE ON HOMES AND HOUSES UNDER CONSTRUCTION AND EXTENSIVE MODERNIZATION SHOULD NEVER BE NEGLECTED. SPECIAL RISK TYPE POLICIES CAN BE SECURED FOR THE PERIOD OF CONSTRUCTION AND CON-VERTED WHEN THE WORK IS COMPLETED. THIS MAY BE CHECKED WITH THE BUILDER AT THE TIME THE WORK IS CONTRACTED. IN VIEW OF RISING COSTS AND INFLATION, THE INSURANCE ON YOUR HOME SHOULD ALWAYS BE UPGRADED TO COVER REPLACEMENT COSTS. ADEQUATE INSURANCE IS RE-QUIRED BY LENDING AGENCIES.

MORTGAGE INSURANCE IS AVAILABLE TO COVER MORTGAGES IN FULL IN CASE OF DEATH OR UNUSUAL MISFORTUNE THAT MIGHT BEFALL MORTGAGEES. SOME LENDING INSTITUTIONS MAY REQUIRE THIS INSURANCE AS PART OF THEIR LENDING AGREEMENT.

HEATING AND AIR CONDITIONING TERMS-----

MANY OF THE DEFINITIONS AND TERMS SHOWN IN THIS LIST WILL APPLY TO BOTH YOUR HEATING AND AIR CONDITIONING PLANNING. IN MOST NEW HOMES AND MANY MODERNIZATION OR RESTORATION PROJECTS TODAY, COMBINATION UNITS ARE BEING INSTALLED WHERE THE CLIMATIC CONDITIONS REQUIRE BOTH CONVENIENCES.

AIR CIRCULATION. THE MOVEMENT OF AIR FROM THE HEATING OR AIR CONDITIONING UNIT THROUGHOUT THE HOME.

AIRTIGHTNESS. THE ABILITY OF THE DUCT WORK TO PREVENT LOSS OF AIR AT THE SEAMS, JOINTS AND CONNECTIONS.

AQUASTAT. A DEVICE TO REGULATE THE TEMPERATURE OF HOT WATER HEAT.

BLOWER. AN ELECTRIC FAN WHICH FORCES AIR THROUGH THE DUCT WORK.

CHRONOTHERM. A COMBINATION CLOCK AND THERMOSTAT SET TO CONTROL HEAT AND COOLING AT PRESET HOURS IF DESIRED.

COMFORT LINE. A LINE OR DISTANCE ABOVE THE FLOOR WHERE ADULTS REQUIRE CONSTANT COMFORTABLE HEATING OR COOLING.

CONVECTOR. AN ELECTRIC OR HOT WATER ROOM HEATING UNIT.

COOLING UNIT. A COOLING COIL INSTALLED IN THE DUCTWORK OF A HEATING SYSTEM FOR SUMMER COOLING.

DUCTWORK. THE PIPING SYSTEM FOR HEATING OR COOLING UNITS. THEY MAY BE ROUND, RECTANGULAR OR SQUARE.

ELECTRONIC CONTROLS. ANY OF THE ELECTRONIC CONTROLS USED IN THE OPERATION OF EITHER HEATING OR COOLING EQUIPMENT.

ELECTRIC HEAT. HEATING BY ELECTRICITY. MAY BE BY SEVERAL METHODS, INCLUDING THE HEAT PUMP.

FILTERS. SCREEN-TYPE UNITS MADE OF SPUN GLASS, OIL TREATED PAPER OR OTHER SYNTHETIC MATERIALS TO FILTER THE AIR FROM HEATING OR COOLING UNITS.

FLOOR REGISTER OR GRILLE. UNITS INSTALLED IN FLOORS OR WALLS TO DISBURSE HEATING OR COOLING INTO THE ROOM.

FORCED AIR. HOT OR COOL AIR THAT IS BLOWN FROM A HEATING OR COOLING UNIT BY MEANS OF A BLOWER.

HEATING OR COOLING SEASON. THE PERIOD OF THE YEAR THAT REQUIRES EITHER HEATING OR COOLING.

HEAT LOSS. TOTAL AMOUNT OF HEAT LOST THROUGH THE WALLS, ROOF, CEILING, FLOORS, OR AROUND WINDOWS OR DOORS.

HEAT PUMPS. UNITS THAT WORK ON THE SAME PRINCIPLE AS REFRIGERATORS. HEAT IS TAKEN FROM THE GROUND THROUGH PIPES OR COILS.

HUMIDITY. THE AMOUNT OF MOISTURE IN THE AIR.

(CONTINUED)

HUMIDIFIER. EQUIPMENT THAT ADDS MOISTURE TO THE AIR—USUALLY IN CONNECTION WITH THE HEATING SYSTEM.

LIMIT CONTROL. A THERMOSTAT THAT PREVENTS OVERHEATING OF A HEATING DEVICE.

PANEL HEATING. A METHOD OF HEATING BY MEANS OF ELECTRICAL OR HOT WATER COILS INSTALLED IN FLOORS, WALLS, CEILINGS.

PERIMETER HEATING. ELECTRIC, HOT AIR OR HOT WATER HEATING INSTALLED AROUND THE ROOM NEAR THE OUTSIDE WALLS.

RADIANT HEATING. ELECTRIC HEATING OF HOMES BY MEANS OF SURFACE RADIATION.

RELATIVE HUMIDITY. THE PERCENTAGE OF MOISTURE IN THE AIR.

SOLAR HEAT. THE METHOD OF USING HEAT OF THE SUN FOR HOME HEATING AND HOT WATER REQUIREMENTS.

SOLAR ORIENTATION. THE DESIGN OF HOMES USING LARGE GLASS AREAS TO PERMIT THE SUN'S HEAT TO ENTER THE AREA.

THERMOSTAT. A DEVICE CONTROLLING THE TEMPERATURE IN A GIVEN AREA.

UNBALANCED SYSTEM. A SYSTEM DESIGNED TO GIVE DIFFERENT DEGREES OF HEATING OR COOLING IN VARIOUS PARTS OF THE HOUSE.

WARM AIR HEATING. ANY HOT AIR HEATING SYSTEM WHERE WARM AIR IS FURNISHED TO THE AREA BY MEANS OF FANS OR GRAVITY.

WET HEAT. HOT WATER OR STEAM SYSTEMS.

WINTER AIR CONDITIONING. GENERALLY USED TO DESCRIBE A FORCED AIR SYSTEM.

ZONE CONTROLS. A SYSTEM USING TWO OR MORE THERMOSTATS OR CHRONOTHERMS.

Notes:

HEATING-----

THE MOST COMMONLY USED HEATING SYSTEM IN HOMES TODAY IS FORCED AIR. THESE HOT AIR HEATING UNITS MAY USE NATURAL OR MANUFACTURED GAS OR OIL AND SOME UTILIZE ELECTRICAL UNITS TO PROVIDE THE HEAT. THEY MAY BE COMBINED WITH COOLING UNITS AND AIR CLEANING TO PROVIDE CENTRAL AIR CONDITIONING USING THE SAME BASIC UNIT AND EQUIPMENT. OTHER TYPES NOW IN USE ARE ELECTRICALLY OPERATED HEAT PUMPS AND HOT WATER HEATERS USING VARIOUS FUELS. HOT WATER UNITS MAY USE BASEBOARD CONVECTORS, RADIATORS OR PIPE EMBEDDED IN THE FLOOR AREAS. ELECTRIC BASEBOARD AND CEILING HEATING UNITS ARE FOUND IN SOME AREAS.

THIS SKETCH SHOWS THE BASIC HOT AIR TYPE HEATING AND AIR CONDITIONING PLANT AND ALL THE COMPONENT PARTS. IT ALSO SHOWS THE ELECTRONIC AIR CLEANER WHICH IS AN OPTIONAL CHOICE ON MOST TYPES.

VARIATIONS IN SIZE, TYPE, OPERATION AND FUEL ARE AVAILABLE FROM VARIOUS MANUFACTURERS.

GENERALLY, THE TYPES OFFERED IN THE DIFFERENT SECTIONS OF THE COUNTRY ARE ADAPTED TO THE CONDITIONS AND CLIMATE PREVALENT IN THAT AREA.

SEE: *HEATING TERMS*.

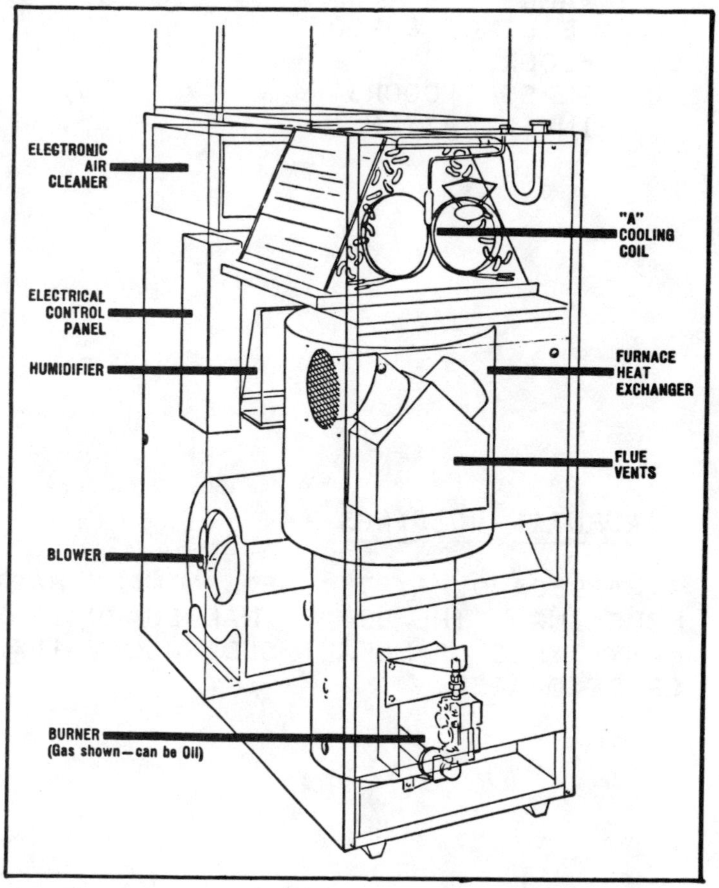

COURTESY U.S. DEPT. OF AGRICULTURE

Notes:

HEAT LOSS, GAIN-----

RESIDENTIAL HEAT LOSSES OR GAINS VARY CONSIDERABLY ACCORDING TO HOUSE CONSTRUCTION, INSULATION, AND ALSO TO A HOME'S LOCATION.

HEAT LOSSES OR GAINS ARE DIFFERENT FOR VARIOUS PARTS OF A HOME. AN APPROXIMATE GUIDE FOR THE PERCENTAGE OF LOSSES/GAINS IS GIVEN IN THE TABLE BELOW. ASSUME A MID-COUNTRY LOCATION (SAY SOUTHERN INDIANA), A TWO STORY HOUSE WITH A BASEMENT.

	LOSS IN WINTER	GAIN IN SUMMER
WALLS	33. %	34%
CEILINGS & ROOF	22.	24
FLOORS	.5	4
GLASS & DOORS	30.	26
INFILTRATION	15.	12

HARDBOARD MOLDINGS-----

HARDBOARD MAY BE USED IN MANY PARTS OF A NEW HOME OR FOR RE-MODELING. THE USE OF HARDBOARD IS SIMPLIFIED BY THE USE OF THE MANY TYPES OF METAL MOLDINGS AVAILABLE. SOME TYPES ARE COLORED OR DECORATED.

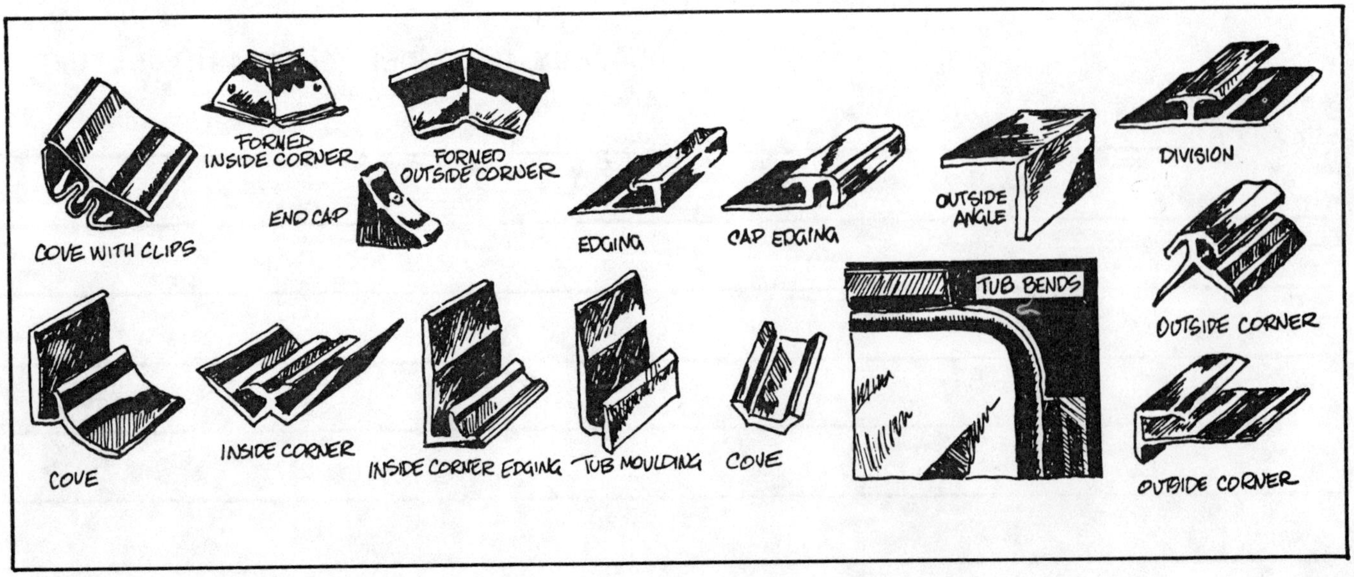

HARDBOARD TERMS-----

HARDBOARD IS A MANUFACTURED PRODUCT MADE FROM WOOD OR OTHER PRODUCTS PRESSED TOGETHER. WHEN YOU DISCUSS OR CONSIDER THIS MATERIAL YOU WILL ENCOUNTER SOME OF THE TERMS LISTED HERE.

EMBOSSED HARDBOARD - A TYPE OF HARDBOARD WITH A DECORATIVE PATTERN IMPRINTED OR PRESSED IN OR ON ITS SURFACE.

FLAKEBOARD - A COMPOSITION BOARD MADE OF SHAVING-LIKE FLAKES OF WOOD THAT IS BONDED AND PRESSED TOGETHER FLAT WITH SYNTHETIC RESINS.

GRAINING - NATURAL WOOD GRAINING PATTERNS PRINTED ON THE SURFACE OF BOARD OR PANEL.

GROOVED HARDBOARD - BOARD THAT HAS REGULAR OR RANDOM SPACED GROOVES THROUGHOUT THE WIDTH OR LENGTH OF THE PANEL.

PARTICLEBOARD - A COMPOSITION BOARD CONSISTING OF DISTINCT PARTICLES OF WOOD BONDED TOGETHER WITH A SYNTHETIC RESIN OR OTHER ADDED BINDER.

PERFORATED HARDBOARD, PEGBOARD - HARDBOARD WITH PUNCHED HOLES SPACED THROUGHOUT THE PANEL. USED FOR HANGING PICTURES AND TOOLS AND ALSO FOR ACOUSTICAL AND DECORATING APPLICATIONS.

SCREEN-BACK HARDBOARD - HARDBOARD WITH ONE SURFACE HAVING A SCREEN-LIKE IMPRESSION.

STANDARD HARDBOARD - REGULAR HARDBOARD AS IT COMES FROM THE PRESS WITHOUT ADDITION OF SPECIAL CHEMICALS OR OTHER SPECIALIZED TREATMENT.

STRIATED HARDBOARD - HARDBOARD THAT HAS A "COMBED" OR CLOSELY GROOVED SURFACE.

S1S HARDBOARD - HARDBOARD HAVING ONE SMOOTH SURFACE AND ONE SCREEN SURFACE.

S2S HARDBOARD - HARDBOARD HAVING TWO SMOOTH SIDES OR SURFACES.

TEMPERED HARDBOARD - HARDBOARD MADE BY ADDING SMALL QUANTITIES OF SPECIAL OILS OR RESINS TO STANDARD HARDBOARD, WHICH IS THEN HEAT-TREATED—OR "TEMPERED"—TO INCREASE STIFFNESS, STRENGTH, MOISTURE RESISTANCE AND ABRASION.

Notes: ______________________________

HARDWARE, UTILITY TYPES-----

HARDWARE OF THIS TYPE IS USED FOR MANY COMMON PURPOSES. THE MANY TYPES AND KINDS SHOWN HERE ARE USUALLY FOUND IN HARDWARE STORES. THEY MAY BE AVAILABLE IN CHROME, BRASS OR ECONOMICAL ZINC COATED TYPES.

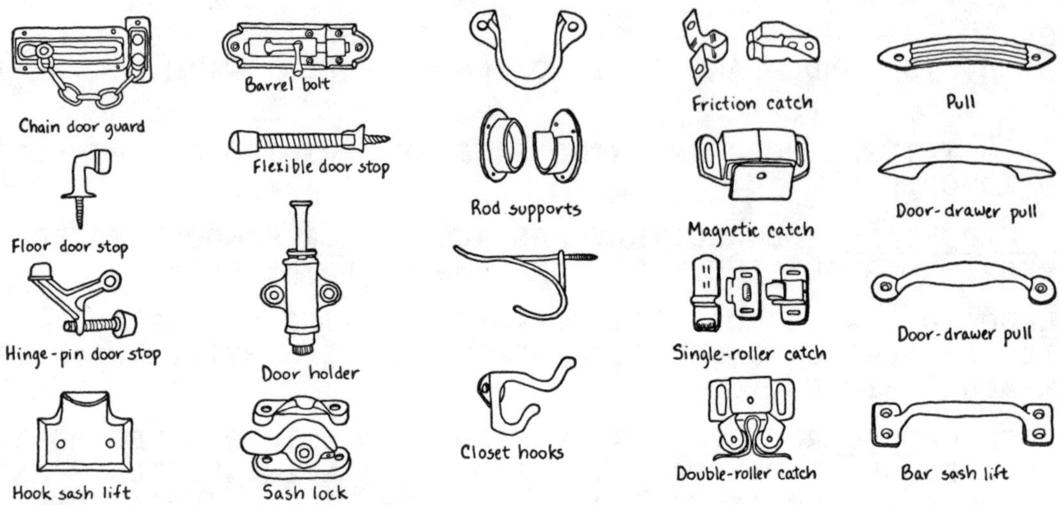

COURTESY SUCCESSFUL HOW TO BUILD YOUR OWN HOME, BY ROBERT C. RESCHKE (STRUCTURES PUBLISHING CO.)

SEE: *LOCKS, KEY PLANS, HINGES.*

HINGES-----

THE SKETCHES SHOWN HERE WILL HELP YOU UNDERSTAND THE VARIOUS TYPES OF HINGES COMMONLY USED IN HOME CONSTRUCTION.

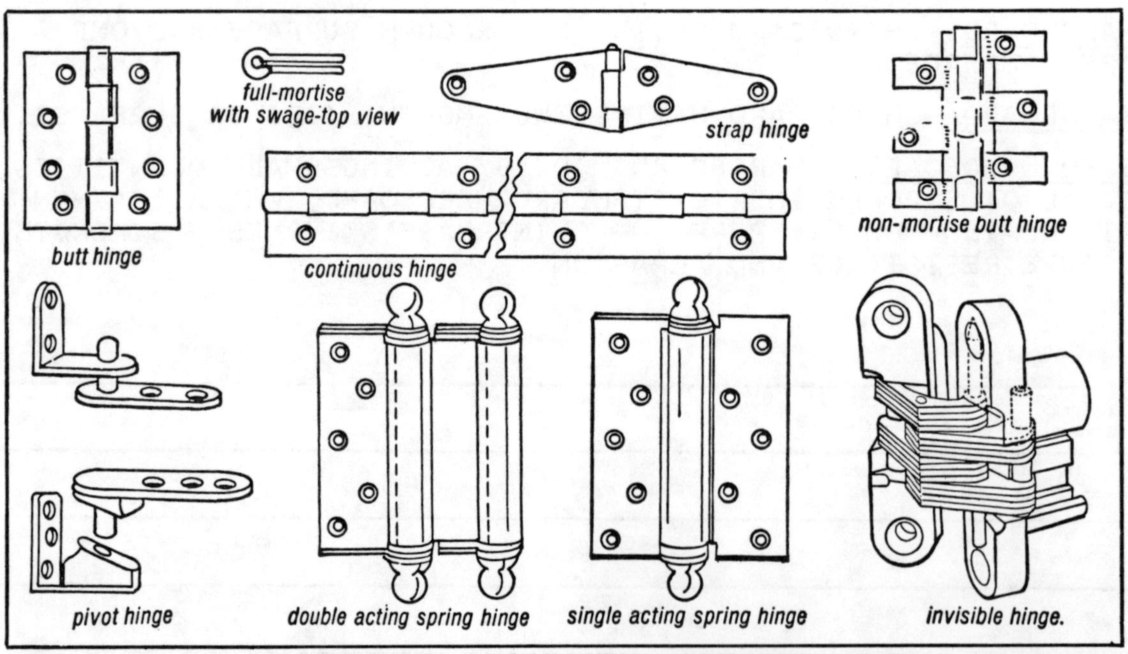

COURTESY SUCCESSFUL PUTTING IT ALL TOGETHER
(STRUCTURES PUBLISHING CO.)

HOUSE CONSTRUCTION, TERMS-----

- BATTEN - THIN STRIPS OVER VERTICAL SIDING JOINTS.
- BOTTOM PLATE - LOWEST HORIZONTAL MEMBER OF A WALL.
- COLLAR BEAM - A TIE BEAM BETWEEN RAFTER PAIRS.
- CRIPPLE - A PART OF A FRAME LESS THAN FULL LENGTH.
- FASCIA - BOARD AT ENDS OF RAFTERS.
- FLOOR FRAMING - KNOWN AS JOISTS.
- HEADER - SUPPORTS FRAMING MEMBERS ACROSS AN OPENING.
- O.C. - ON CENTER. FOR MEASUREMENT.
- RAKE - ROOF OVERHANG AT GABLE END.
- RIDGE BOARD - BOARD AT PEAK OF ROOF.
- ROOF FRAMING - KNOWN AS RAFTERS.
- SILL - ALSO MUDSILL OR PLATE, THE LOWEST FRAMING MEMBER OF A BUILDING.
- SOFFIT - UNDERSIDE OF THE ROOF OVERHANG.
- SPAN - THE DISTANCE BETWEEN THE SUPPORTS OF A FRAMING MEMBER.
- STUDS - VERTICAL MEMBERS OF WALL.
- SUBFLOORING - BOTTOM LAYER OF A TWO LAYER FLOOR.
- TOP PLATE - TOP MEMBER OF FRAME.
- WALL FRAMING - KNOWN AS STUDS.
- UNDERLAYMENT - SMOOTH WOOD BASE SHEETS APPLIED UNDER RESILIENT FLOORING OR CARPET.

INSULATION, ENERGY EFFICIENCY, "R" RESISTANCE-----

THE ABILITY OF A BUILDING MATERIAL OR PRODUCT TO INSULATE AGAINST HEAT AND COLD IS MEASURED BY THE "R" (RESISTANCE VALUE). INSTALLED RESISTANCE (R) IS THE SUM OF THE RESISTANCE OF ALL MATERIALS AND PRODUCTS USED, INCLUDING THE AIR SPACE AND SURFACE. THE HIGHER THE R-VALUE, THE MORE EFFICIENT THE INSULATION IS RETARDING THE PASSAGE OF HEAT OR COLD.

Notes: ______________________________

____

____

____

____

____

HOUSE PARTS------

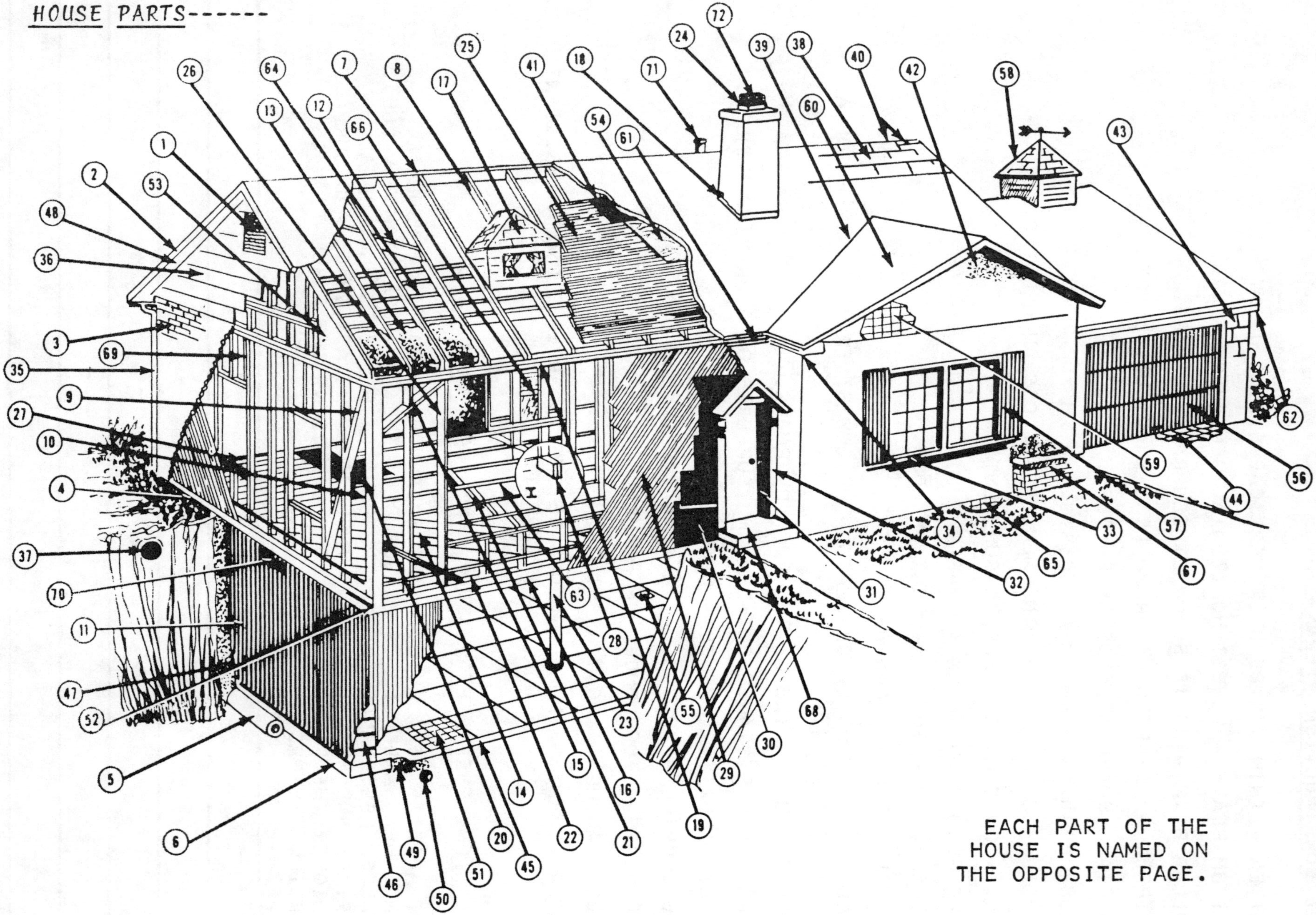

EACH PART OF THE
HOUSE IS NAMED ON
THE OPPOSITE PAGE.

SEE: *DOOR TYPES, WINDOW TYPES, FIREPLACE PARTS, CHIMNEY PARTS, ROOF TYPES, FOUNDATIONS, FRAMING DETAILS, GARAGE SIZES.*

HOUSE PARTS-----

USE THIS HANDY LIST TO LOCATE THE VARIOUS HOUSE PARTS SHOWN ON THE OPPOSITE PAGE.

1. LOUVER
2. EAVE
3. BRICK VENEER
4. GRADE LINE
5. DRAIN TILE
6. FOOTING
7. RIDGE RAFTER
8. RAFTER
9. DIAGONAL BRACING
10. CORNER POST
11. CEMENT PLASTER (PARGETING) AND FOUNDATION COATING
12. COLLAR BEAM
13. INSULATION
14. SUB-FLOORING
15. KNEE BRACING
16. SILL
17. DORMER (WITH HIP ROOF)
18. FLASHING
19. PLATE
20. FLOORING PAPER
21. CROSS BRIDGING
22. FIRE STOP
23. LALLY (METAL) COLUMN
24. CHIMNEY & FLUE LINER
25. ROOF SHEATHING
26. STUDDING
27. FINISH FLOORING
28. GIRDER OR "I" BEAM
29. SHEATHING
30. BUILDING PAPER OR VAPOR SEAL
31. DOOR JAMB
32. PILASTER
33. WINDOW SILL
34. GUTTER
35. DOWNSPOUT OR LEADER
36. SIDING
37. STORM SEWER
38. SHINGLES
39. VALLEY
40. RIDGE
41. ROOFING FELT
42. STUCCO
43. WOOD SHINGLES
44. REINFORCING WIRE
45. CONCRETE FLOOR
46. MASONRY OR CONCRETE FOUNDATION
47. GRAVEL
48. SOFFIT
49. FILL
50. SANITARY SEWER
51. FLOOR TILE
52. SILL SEALER
53. SHEATHING BOARD
54. PLYWOOD ROOF SHEETING
55. BELL TRAP
56. OVERHEAD GARAGE DOOR
57. SHUTTER
58. CUPOLA
59. STUCCO MESH OR LATH
60. GABLE ROOF
61. CORNICE
62. FRIEZE BOARD
63. FLOOR JOIST
64. CEILING JOIST
65. AREA WALL
66. LATH AND PLASTER
67. PLANTER
68. STOOP
69. CRIPPLE
70. FOUNDATION GRATE
71. SOIL STACK
72. CHIMNEY SCREEN

Notes: ____________________

INSULATION STANDARDS-----

AT THIS PRESENT WRITING, THERE'S DISAGREEMENT ABOUT WHAT STANDARDS OF INSULATION SHOULD BE PROVIDED AND BY WHOM. SOME STATE ENERGY AGENCIES HAVE ESTABLISHED SUCH STANDARDS. PROBABLY THE MOST EXTENSIVE WORK IN THIS FIELD TO DATE HAS BEEN THAT LEADING TO THE ASHRAE-75 STANDARD ISSUED BY THE AMERICAN SOCIETY OF HEATING, REFRIGERATION AND AIR CONDITIONING ENGINEERS.

MUCH DISCUSSION AND ARGUMENT IS PRESENTLY OCCURRING ON A NATIONAL SCALE ABOUT THE BUILDING ENERGY PERFORMANCE STANDARDS PROPOSED BY THE U.S. DEPARTMENT OF ENERGY AS A RESULT OF A MANDATE FROM CONGRESS. UNTIL SUCH TIME IN THE FUTURE WHEN THESE STANDARDS ARE ISSUED AND BECOME EFFECTIVE (OR THEIR STATUS OTHERWISE RESOLVED), IT IS RECOMMENDED THAT GUIDANCE IN HOUSING INSULATION PRACTICES BE IN ACCORDANCE WITH THE PUBLICATION DESCRIBED BELOW.

THE NATIONAL ASSOCIATION OF HOME BUILDERS HAVE ISSUED A 20 PAGE BOOKLET ENTITLED "NAHB THERMAL PERFORMANCE GUIDELINES FOR ONE AND TWO FAMILY DWELLINGS". THE GUIDELINES ARE PRACTICAL, COMPREHENSIVE AND FLEXIBLE. IN ADDITION, THEIR PRESENTATION IN THE BOOKLET IS EASY TO UNDERSTAND AND COVERS THE ENTIRE TOPIC IN A WELL ORGANIZED MANNER. TECHNICAL TERMS ARE EXPLAINED AND AN EXAMPLE IS GIVEN OF HOW TO USE THE GUIDELINES. IN ADDITION, WORKSHEETS ARE PROVIDED TO HELP IN CALCULATING KEY INDICES AND ENERGY CONSERVING TECHNIQUES. THE BOOKLET IS OBTAINABLE FROM NAHB, 15TH & M STS., WASHINGTON, DC 20005.

SEE: *ENERGY, INSULATION, ENERGY EFFICIENCY, "R" RESISTANCE.*

INSULATION TYPES-----

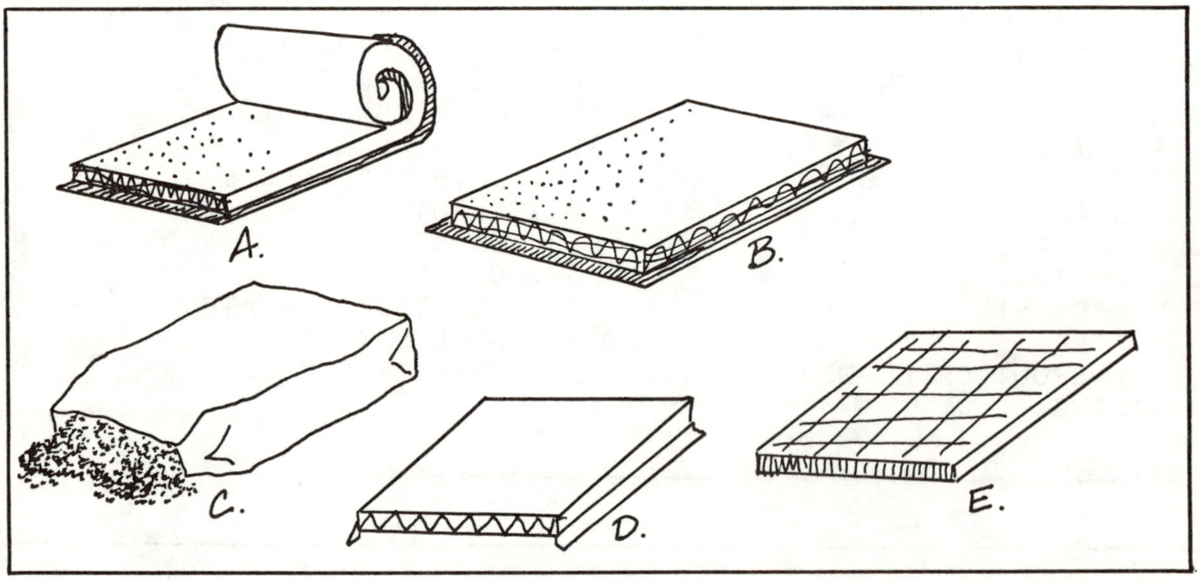

A. ROLL OR BLANKET TYPE WITH VAPOR BARRIER.
B. BATT TYPE WITH VAPOR BARRIER.
C. LOOSE FILL TYPE COMES IN BAGS.
D. REFLECTIVE TYPE.
E. RIGID TYPE, BOARD.

INSULATION AND YOUR HOME-----

THIS CUTAWAY SECTION OF AN AVERAGE HOUSE WILL SHOW YOU WHERE INSULATION MAY BE PLACED TO PREVENT THE LOSS OF HEAT IN COLD WEATHER AND PROTECTION FROM HEAT IN THE SUMMER.

GOOD QUALITY, PROPERLY PLACED INSULATION OF THE PROPER TYPE AND INSULATION VALUE HAS ONE BASIC FUNCTION: THAT IS TO PROVIDE COMFORTABLE LIVING IN BOTH COLD AND HOT WEATHER WITH A MINIMUM AMOUNT OF EXPENDED ENERGY. WITH STRONG EMPHASIS ON ENERGY CONSERVATION AND THE COST OF ENERGY, EFFICIENT INSULATION IS A VERY IMPORTANT FACTOR IN BUILDING, BUYING, REMODELING, OR RESTORING.

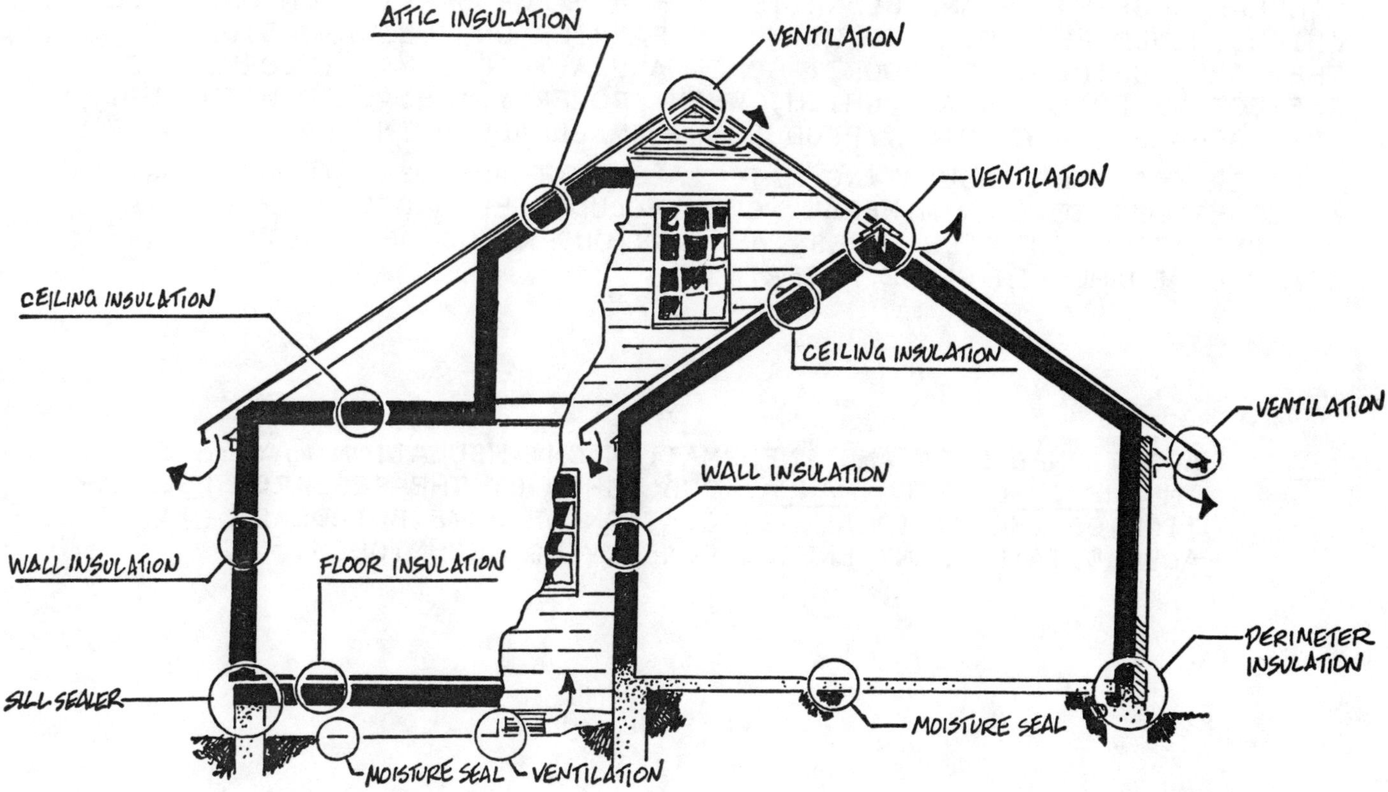

Notes: ______

INSULATION MATERIALS-----

THE MOST COMMON RESIDENTIAL TYPE OF INSULATION IN USE TODAY IS MADE OF FIBERGLASS. IT IS AVAILABLE IN BATT, BLANKET, AND LOOSE FILL TYPES. IT ALSO COMES IN AN INSULATING SHEATHING FORM.

FIBERGLASS INSULATING MATERIALS ARE PART OF A FAMILY OF INSULATIONS IN THE MINERAL WOOL CATEGORY. LESS COMMONLY USED MINERAL WOOL FIBROUS MATERIALS ARE MADE FROM MOLTEN ROCK OR SLAG. ALL TYPES ARE CLASSED AS INCOMBUSTIBLE. IN ADDITION TO THE ABOVE TYPES, SOME MANUFACTURERS CAN FURNISH A BLOWING WOOL DESIGNED FOR PNEUMATIC APPLICATION.

A COMMON PRACTICE IS FOR THE INSULATION MANUFACTURER TO FURNISH MINERAL WOOL BATTS AND BLANKETS WITH A VAPOR BARRIER ON ONE SIDE WHICH SHOULD BE INSTALLED WITH THE BARRIER ON THE WARM SIDE OF THE WALL, CEILING OR FLOOR. SOME MANUFACTURERS MAY INCORPORATE A REFLECTIVE FOIL SURFACE WHICH, WHEN PROPERLY INSTALLED WITH SUITABLE ADJACENT AIR SPACE, PRODUCES A SMALL ADDED INSULATIVE VALUE.

OTHER FORMS OF INSULATING MATERIALS THAT ARE USED IN HOUSING TO A LESSER DEGREE THAN MINERAL WOOL INCLUDE CELLULOSE VERMICULITE AND PERLITE LOOSE FILL TYPES, AND VARIOUS KINDS OF PLASTIC FOAMBOARD OR FOAMED-IN-PLACE MATERIALS.

NOTE: ADDITIONAL DATA AND INFORMATION ON INSULATION MAY BE FOUND IN THE INSULATION MANUAL PUBLISHED BY THE RESEARCH FOUNDATION OF THE NATIONAL ASSOCIATION OF HOME BUILDERS. IT ALSO DETAILS MANY ENERGY CONSERVING TECHNIQUES.

Notes: ______

INTEREST RATES-----

YOU CAN USE THIS SIMPLE TABLE OF AMORTIZATION RATES TO CALCULATE THE MONTHLY INTEREST ON ANY SIZE MORTGAGE OR LOAN.

NUMBER OF YEARS	INTEREST RATES 8%	9%	10%	11%	12%	13%	14%	15%	16%
3	$3.13	$3.18	$3.23	$3.28	$3.33	$3.37	$3.42	$3.47	$3.52
4	2.44	2.49	2.54	2.59	2.64	2.70	2.74	2.79	2.84
5	2.02	2.07	2.12	2.18	2.23	2.28	2.33	2.38	2.44
6	1.75	1.80	1.85	1.91	1.96	2.01	2.07	2.12	2.17
7	1.56	1.61	1.66	1.72	1.77	1.82	1.88	1.93	1.99
8	1.39	1.44	1.52	1.58	1.63	1.69	1.74	1.80	1.86
9	1.29	1.35	1.41	1.47	1.52	1.58	1.64	1.70	1.76
10	1.20	1.26	1.32	1.38	1.44	1.50	1.56	1.62	1.68
11	1.12	1.18	1.25	1.31	1.37	1.43	1.49	1.56	1.62
12	1.07	1.13	1.20	1.26	1.32	1.38	1.44	1.51	1.57
13	1.02	1.08	1.15	1.21	1.27	1.34	1.40	1.47	1.53
14	.98	1.04	1.11	1.17	1.24	1.30	1.37	1.43	1.50
15	.94	1.01	1.07	1.14	1.21	1.27	1.34	1.40	1.47
16	.92	.98	1.05	1.11	1.18	1.24	1.31	1.38	1.45
17	.89	.95	1.02	1.09	1.16	1.22	1.29	1.36	1.43
18	.86	.93	1.00	1.07	1.14	1.21	1.28	1.35	1.42
19	.84	.91	.98	1.05	1.12	1.19	1.26	1.33	1.41
20	.83	.90	.97	1.04	1.11	1.18	1.25	1.32	1.40
25	.77	.84	.91	.99	1.06	1.13	1.21	1.29	1.36
30	.72	.80	.88	.96	1.03	1.11	1.19	1.27	1.35

TO CALCULATE THE MONTHLY INTEREST CHARGE, FIND THE NUMBER WHERE THE INTEREST RATE AND THE NUMBER OF YEARS THE LOAN WILL BE FINANCED INTERSECT ON THE CHART. MULTIPLY THE AMOUNT OF THE LOAN BY THIS FIGURE.

FOR EXAMPLE, A LOAN OF $3300 FOR A 12 YEAR PERIOD AT 12 PERCENT INTEREST WOULD BE $1.32 MULTIPLIED BY $3300 OR APPROXIMATELY $43.56 PER MONTH.

Notes: ______

INTEREST-----

RATE COMPARISONS MAY BE DETERMINED BY USING THIS TABLE.

RATE CHARGED (ADD-ON TYPE)	SIMPLE ANNUAL RATE
$ 4 PER $100, OR 4% A YEAR	7.4%
$ 6 PER $100, OR 6% A YEAR	11.1%
$ 8 PER $100, OR 8% A YEAR	14.8%
$10 PER $100, OR 10% A YEAR	22.2%
3/4 OF 1% A MONTH ON UNPAID BALANCE	9 %

RATE CHARGED	SIMPLE ANNUAL RATE
5/6 OF 1% A MONTH ON UNPAID BALANCE	10%
1% A MONTH ON UNPAID BALANCE	12%
1-1/4% A MONTH ON UNPAID BALANCE	15%
1-1/2% A MONTH ON UNPAID BALANCE	18%
2-1/2% A MONTH ON UNPAID BALANCE	30%

LOAN APPLICATIONS-----

THE APPLICATION FOR A HOME OR MODERNIZATION LOAN IS A RELATIVELY SIMPLE MATTER AND TAKES VERY LITTLE TIME. THE BUYER OR REALTOR SUBMITS HIS SALES AGREEMENT, OR IN THE CASE OF A NEW HOME THE PLANS, SPECIFICATIONS, COST ESTIMATE FOR THE APPRAISAL AND EVALUATION, TO THE LENDING AGENCY. THE BUYER IS QUESTIONED REGARDING HIS AGE, MARITAL STATUS, FAMILY, EMPLOYMENT, INCOME, OBLIGATIONS, INSURANCE AND ABILITY TO PAY THE SUGGESTED MONTHLY PAYMENTS. CREDIT REFERENCES ARE REQUIRED FROM MERCHANTS AND OTHER BUSINESS CONCERNS. FOLLOWING THIS, AN APPRAISAL IS MADE OF THE PROPERTY, PROPOSED BUILDING OR PLANNED IMPROVEMENTS TO DETERMINE THE PRESENT OR TOTAL VALUE. BASED ON THESE FINDINGS, USUALLY REQUIRING A WEEK OR TWO, A LOAN IS SUGGESTED BASED ON THE EVALUATION OF THE PROPOSAL. THE APPLICATION COST IS NOMINAL AND IN SOME CASES IT IS FREE TO SINCERE AND LEGITIMATE APPLICANTS. THE LOAN FEE, INTEREST RATES, MONTHLY PAYMENTS, ETC., ARE DETERMINED AT THIS TIME. THE APPLICANT IS NOT OBLIGATED UNTIL ALL THE LEGAL PAPERS ARE PRESENTED AND EXECUTED BY THE PROPER PARTIES AT THE TIME OF CLOSING.

TRUTH IN LENDING----

THIS GOVERNMENT LAW REQUIRES ALL LENDERS TO DISCLOSE TO BORROWERS THE ACTUAL PERCENTAGE RATE CHARGED ON ALL LOANS. WHEN YOU DECIDE ON A LENDING AGENCY TO FINANCE YOUR HOME, MODERNIZATION OR RESTORATION, CAREFULLY CHECK THE VARIOUS TYPES AND METHODS OF FINANCING AND SELECT THE ONE THAT BEST FITS YOUR CONDITIONS.

KITCHEN RANGE CLEARANCES-----

THE KITCHEN RANGE OR COOKING CENTER IS THE FOCAL OR CENTRAL PART OF THE MODERN KITCHEN. THE LOCATION OF THIS EQUIPMENT IS VERY IMPORTANT AS INDICATED BY THE *KITCHEN BASIC WORK TRIANGLE* DATA IN THIS BOOK.

THE RANGE, COOKTOP, CABINET AND RANGE HOOD HEIGHTS AND CLEARANCE ARE VERY IMPORTANT AND SHOULD BE CAREFULLY CONSIDERED.

DIMENSIONS INDICATED ARE FOR AVERAGE HEIGHT PERSONS. IF THEY ARE ADJUSTED FOR SHORT OR TALL PEOPLE IT SHOULD BE REMEMBERED THIS MAY PRESENT A PROBLEM IF THE PROPERTY IS TO BE SOLD AT A LATER DATE.

AS SHOWN IN THE SKETCHES BELOW, THE VARIOUS DIMENSIONS, SPACING AND PLACEMENT ARE CLEARLY INDICATED TO ASSIST YOU IN PLANNING THIS PART OF YOUR KITCHEN DESIGN.

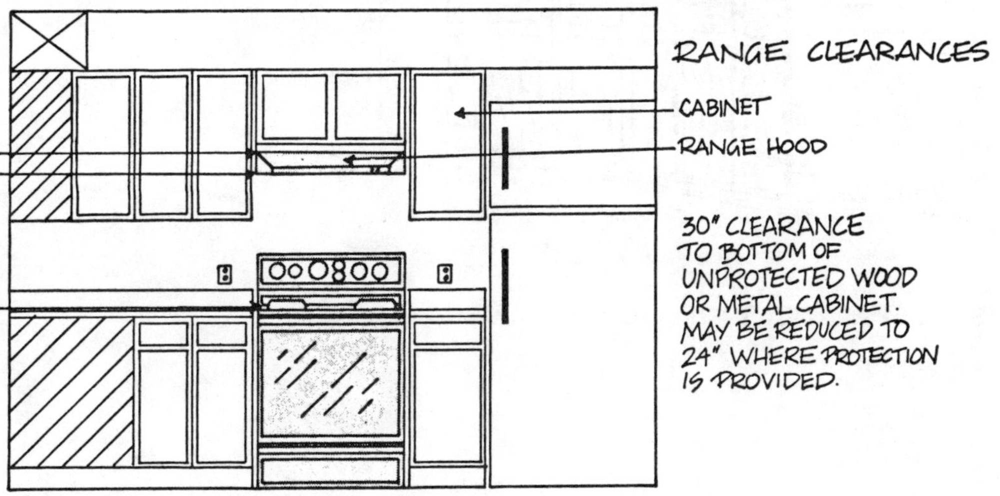

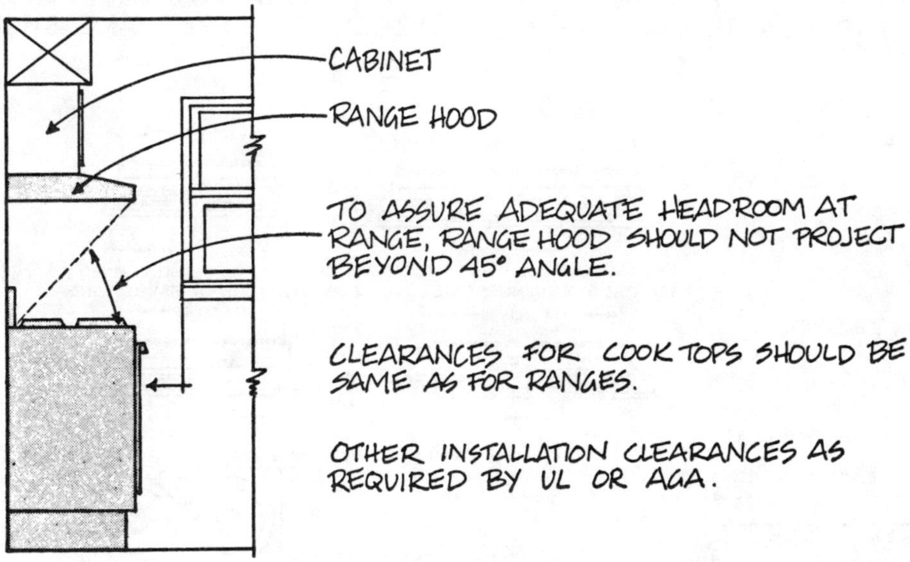

SEE: *KITCHEN WORK HEIGHTS*, *KITCHEN DESIGN DATA*. ALSO, SUCCESSFUL KITCHENS (STRUCTURES PUBLISHING COMPANY)

KITCHEN CABINET PARTS-----

HERE YOU WILL FIND THE LOCATIONS AND NAMES OF KITCHEN CABINET PARTS.

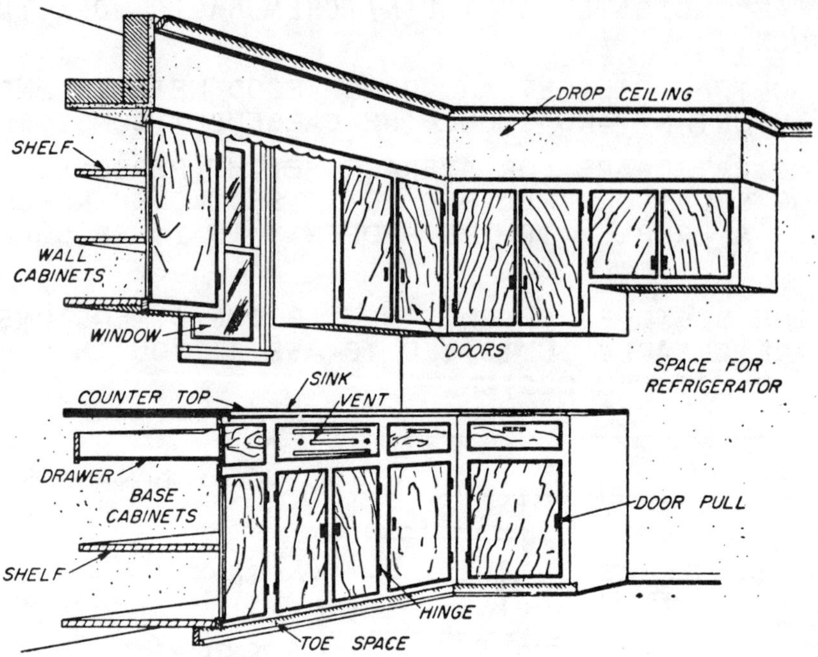

COURTESY OF U.S. DEPARTMENT OF AGRICULTURE.

KITCHEN SINKS-----

THE SIZE, TYPE AND LOCATION OF THE SINKS IN THE KITCHEN LAYOUT IS VERY IMPORTANT. A GROUP OF SIZES AND TYPES ARE SHOWN IN THESE SKETCHES.

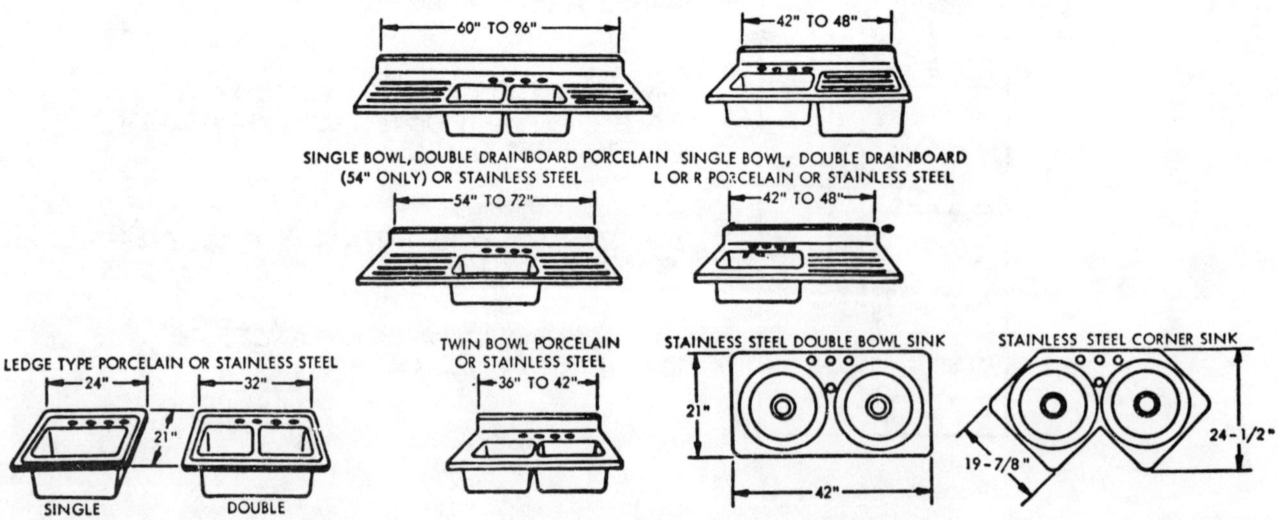

KITCHEN TABLE SEATING-----

MANY MODERN KITCHENS PROVIDE SPACE FOR INFORMAL DINING. THE DIMENSIONS AND SEATING ACCOMMODATIONS SHOWN HERE WILL HELP YOU PLAN YOUR KITCHEN SPACE TO ACCOMMODATE DINING.

RECTANGULAR	
24" X 30"	2 PERSONS
24" X 36"	3 TO 4 PERSONS
44" X 48"	6 TO 8 PERSONS
36" X 60"	6 TO 8 PERSONS

ROUND	
3' DIAMETER	4 PERSONS
4' DIAMETER	5 PERSONS
4-1/2' DIA.	6 PERSONS
5' DIAMETER	7 PERSONS
5-1/2' DIA.	8 TO 10 PERSONS

THIS SKETCH OF KITCHEN TABLES WILL GIVE YOU AN IDEA OF SEATING ARRANGEMENTS. BY USING THE CHART ABOVE YOU CAN EASILY DETERMINE YOUR SPACE REQUIREMENTS.

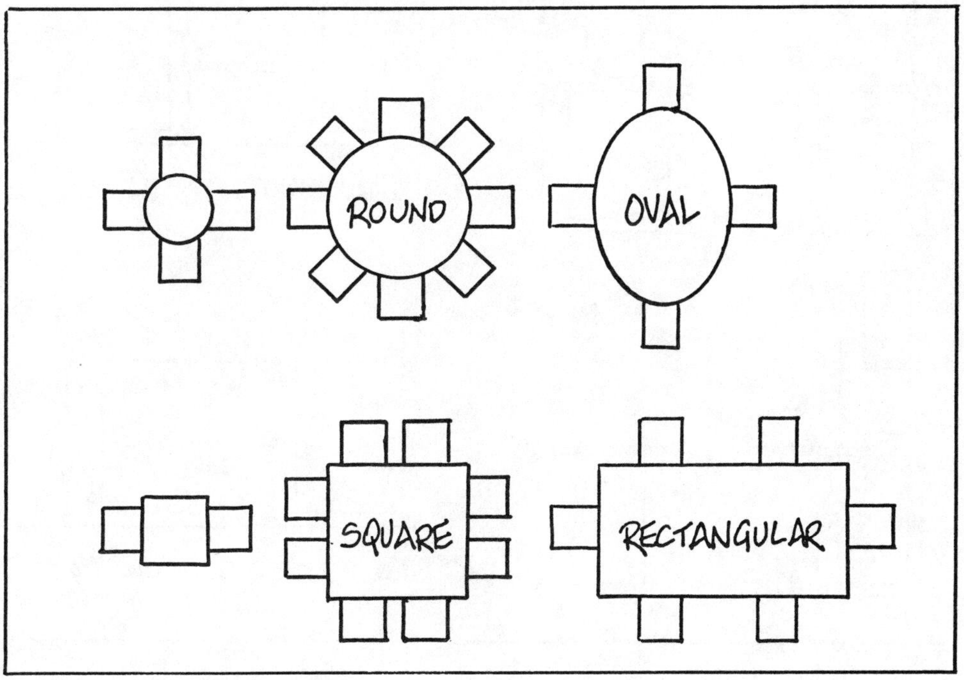

Notes: ______________________________

KITCHEN WORK HEIGHTS-----

THE MODERN KITCHEN IS THE CENTRAL PART OF THE HOME. PRACTICAL AND THOUGHTFUL PLANNING WILL PAY OFF IN EFFICIENCY.

THE BASIC ARRANGEMENTS SHOWN HERE, AS WELL AS THE DIMENSIONS FOR CABINETS, MAY BE CONVERTED TO ANY MODERN SET-UP YOU PLAN.

USE THE SKETCH SHEETS FOUND IN THIS BOOK OR A KITCHEN TEMPLATE TO PLAN YOUR KITCHEN.

THE SKETCHES BELOW SHOW THE AVERAGE HEIGHTS FOR COUNTERS, CABINETS AND CUPBOARDS.

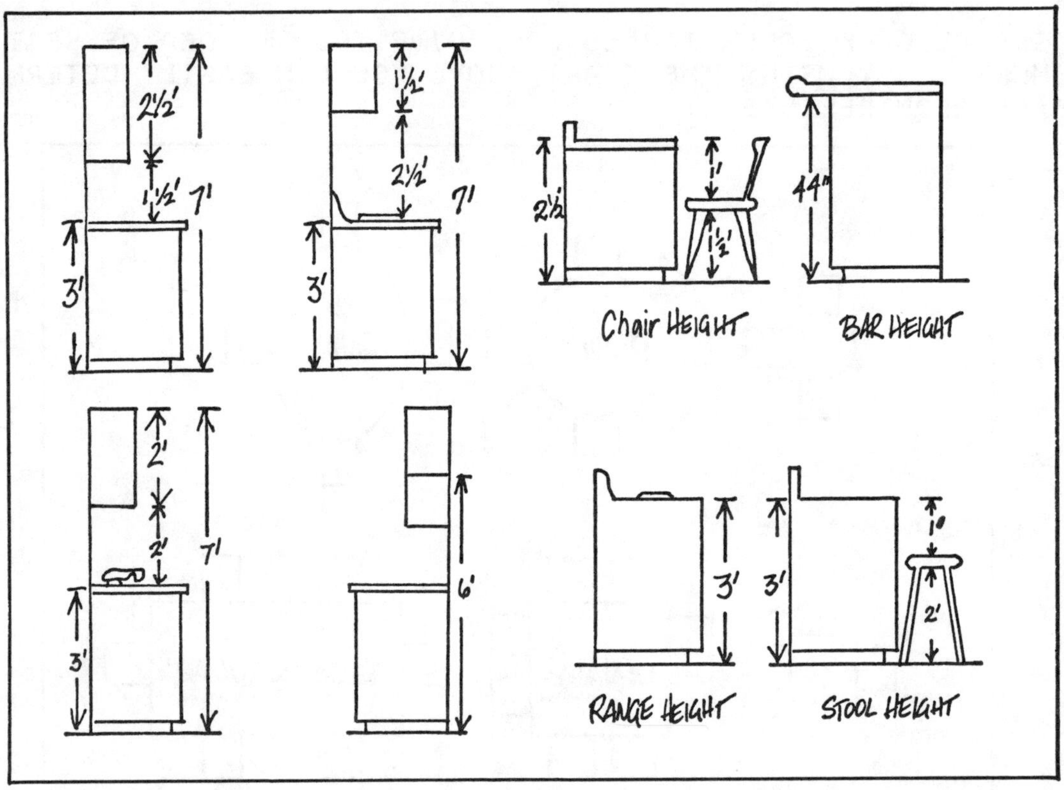

Notes: ______________________________

KITCHEN, BASIC WORK TRIANGLE-----

NEW AND REMODELED KITCHENS SHOULD BE PLANNED BY USING THE WIDELY RECOMMENDED AND PRACTICAL WORK TRIANGLE SHOWN HERE. THIS METHOD CONNECTS THE PREPARATION, REFRIGERATION, COOKING AND CLEAN UP AREAS WITH A MINIMUM OF SPACE REQUIREMENTS.

THE BEST AND MOST PRACTICAL ARRANGEMENTS SUGGEST A TOTAL TRIANGLE DIMENSION OF 12 TO 23 FEET.

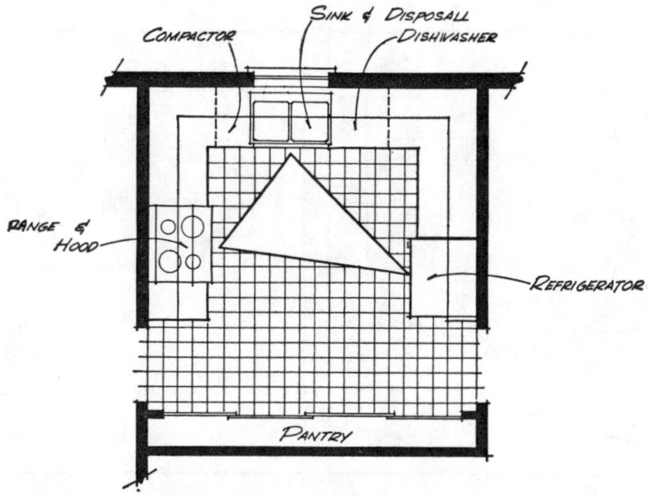

THE SKETCHES BELOW SHOW A NUMBER OF PLANS DEVELOPED AROUND THE TRIANGLE METHOD.

THE IDEAS PRESENTED IN THESE SKETCHES WILL SERVE TO GUIDE YOU IN COORDINATING YOUR IDEAS INTO YOUR PLANNING. USE THE SKETCH SHEETS FOUND IN THIS BOOK TO ASSIST YOU IN DESIGNING AND PLANNING YOUR KITCHEN.

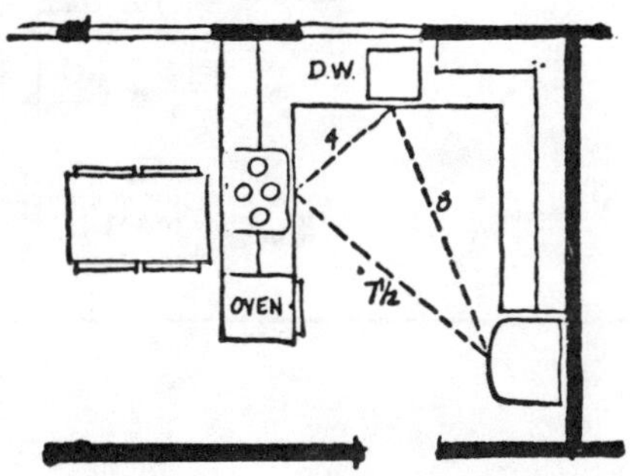

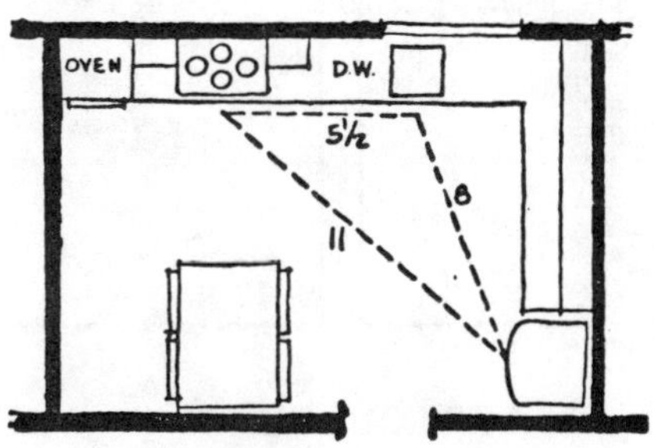

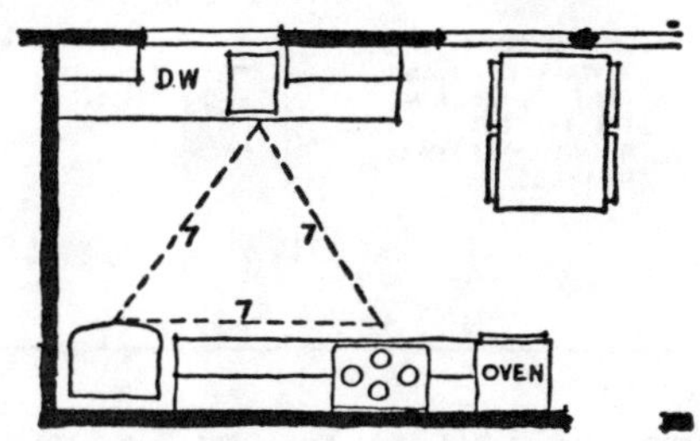

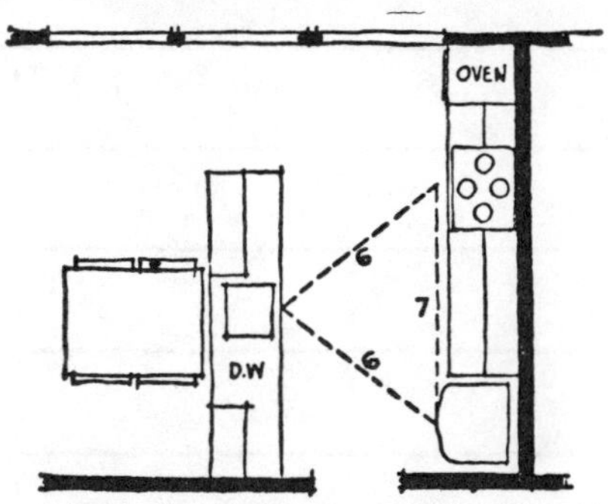

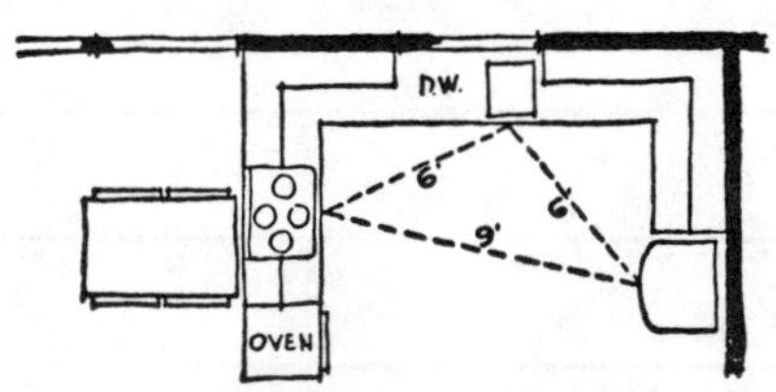

Courtesy U.S. Dep't. Agr.

SEE: *KITCHEN DESIGN DATA, KITCHEN CABINET PARTS, KITCHEN WORK HEIGHTS, KITCHEN SINKS.*

ALSO: SUCCESSFUL KITCHENS (STRUCTURES PUBLISHING CO.)

KITCHEN DESIGN DATA-----

BASIC ARRANGEMENTS MAY BE CONVERTED TO ANY TYPE AND STYLE OF KITCHENS BY USING AND INCORPORATING STANDARD TYPE AND STYLE BASES, CABINETS, REFRIGERATORS, RANGES AND OTHER APPLIANCES IN YOUR PLANS. THE UNUSUAL LAYOUTS THAT MAY BE DEVELOPED WILL DEPEND ON YOUR ABILITY TO VISUALIZE AND PLAN TO SUIT YOUR NEEDS.

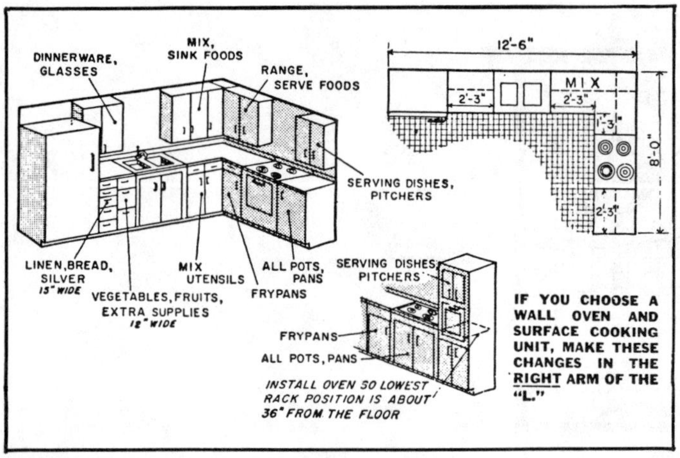

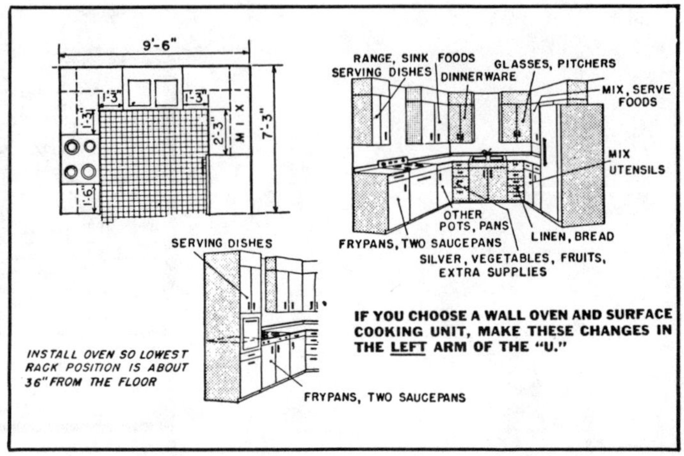

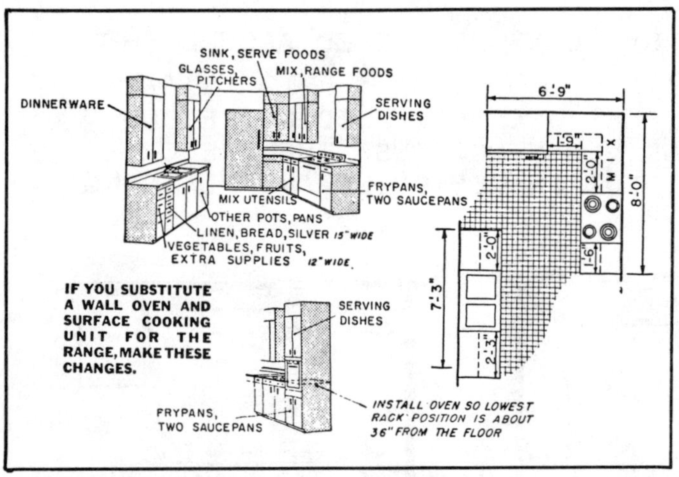

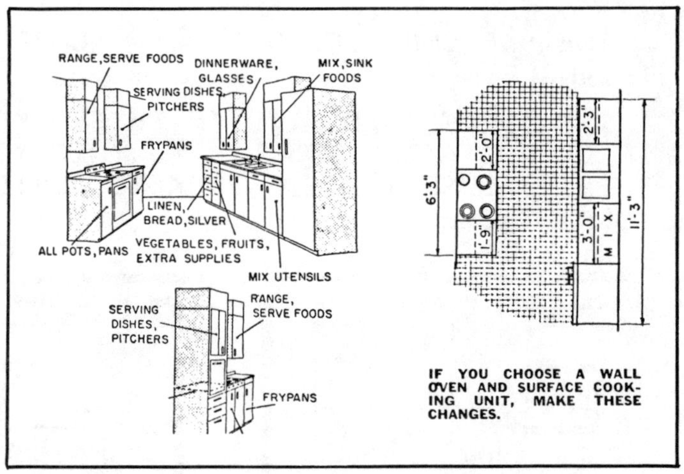

COURTESY U.S. DEPARTMENT OF AGRICULTURE

Notes: ______________________________

____

____

____

____

____

____

LUMBER, MARKING-----

YOU WILL FIND MANY GRADES IN ALL SPECIES OF LUMBER AND IN VARIOUS TYPES, SIZES AND PATTERNS. CERTAIN SPECIES ARE ESPECIALLY ADAPTABLE TO SPECIFIC USES—FROM THE STANDPOINT OF WORKABILITY, COST, UPKEEP, APPEARANCE AND WEARING OR WEATHERING ABILITY. CYPRESS, REDWOOD AND CEDAR, FOR EXAMPLE, ARE PARTICULARLY ADAPTED TO OUTSIDE USES WHERE EXPOSURE TO THE ELEMENTS AND APPEARANCE ARE CONCERNED.

THE VARIOUS PINE SPECIES HAVE UNLIMITED USES FOR MANY PURPOSES DUE TO THEIR AVAILABILITY IN MOST AREAS AND THEIR EASY WORKING QUALITIES. THEY MAY BE TREATED OR PRESERVED IF REQUIRED.

THE HARDWOODS HAVE SPECIFIC USES IN BUILDING AND MAY BE UTILIZED FOR FLOORS, TRIM AND OTHER SPECIAL ADAPTATIONS. ALL TYPES OF WOOD ARE USED IN PLYWOOD MANUFACTURE SO THE COLORS, GRAINING AND TEXTURE MAY BE USED IN ANY PART OF HOME CONSTRUCTION.

PRACTICALLY ALL LUMBER PRODUCED NOW FOR HOME CONSTRUCTION IS GRADED AND MARKED AT THE MILL. THE GRADE MARKINGS ARE USUALLY FOUND ON EACH PIECE. THESE MARKINGS ARE UNDER THE JURISDICTION OF VARIOUS ASSOCIATIONS AND RIGID INSPECTION OF THE VARIOUS BUREAUS.

THE GRADING REGULATIONS ARE USED BY THE ASSOCIATIONS TO ASSURE UNIFORMITY IN QUALITY AND TO MAKE SURE THE PRODUCTS ARE USED PROPERLY TO MEET BUILDING REQUIREMENTS.

THE MARKINGS USUALLY SHOW THE NAME OF THE SPECIES, THE GRADE, CODE NUMBER OF THE PRODUCING MILL, MOISTURE CONTENT AND IDENTIFIES THE ASSOCIATION. OTHER ITEMS MAY BE SHOWN ON SOME TYPES OF LUMBER.

A TYPICAL GRADE MARKING IS SHOWN HERE WITH EXPLANATION OF THE VARIOUS SYMBOLS.

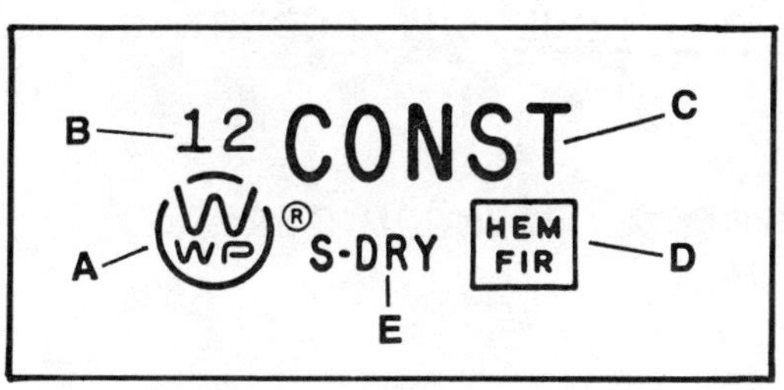

A. MARK OF THE ASSOCIATION
B. MILL OR PRODUCER NUMBER
C. OFFICIAL GRADE NAME ABBREVIATION
D. WOOD SPECIES
E. MOISTURE CONTENT.

SEE: *LUMBER GRADING, LUMBER SIZES, LUMBER KINDS, LUMBER PRICING, LUMBER ABBREVIATIONS, BOARD FEET.*

LUMBER, PLYWOOD GRADE MARKS-----

THESE SAMPLE MARKINGS OR STAMPINGS USED ON LUMBER AND PLYWOOD WILL SERVE TO ACQUAINT YOU WITH SOME OF THE COMMONLY USED MARKS AND GRADES. GOVERNMENT REGULATIONS REQUIRE THESE PRODUCTS BE STAMPED BEFORE SHIPMENT.

BY CHECKING THE VARIOUS LUMBER AND PLYWOOD MARKINGS YOU CAN LOCATE AND PURCHASE THE PRODUCTS BEST SUITED FOR YOUR USES AND ELIMINATE MANY OF THE PROBLEMS OF QUALITY SELECTION.

THIS GROUP SHOWS FOUR WEST COAST LUMBER BUREAU MARKINGS.

TWO PLYWOOD MARKINGS SHOWING INTERIOR AND EXTERIOR GRADES ARE INDICATED BELOW.

FOUR MARKINGS OF THE WESTERN WOOD PRODUCTS ASSOC. ARE SHOWN HERE.

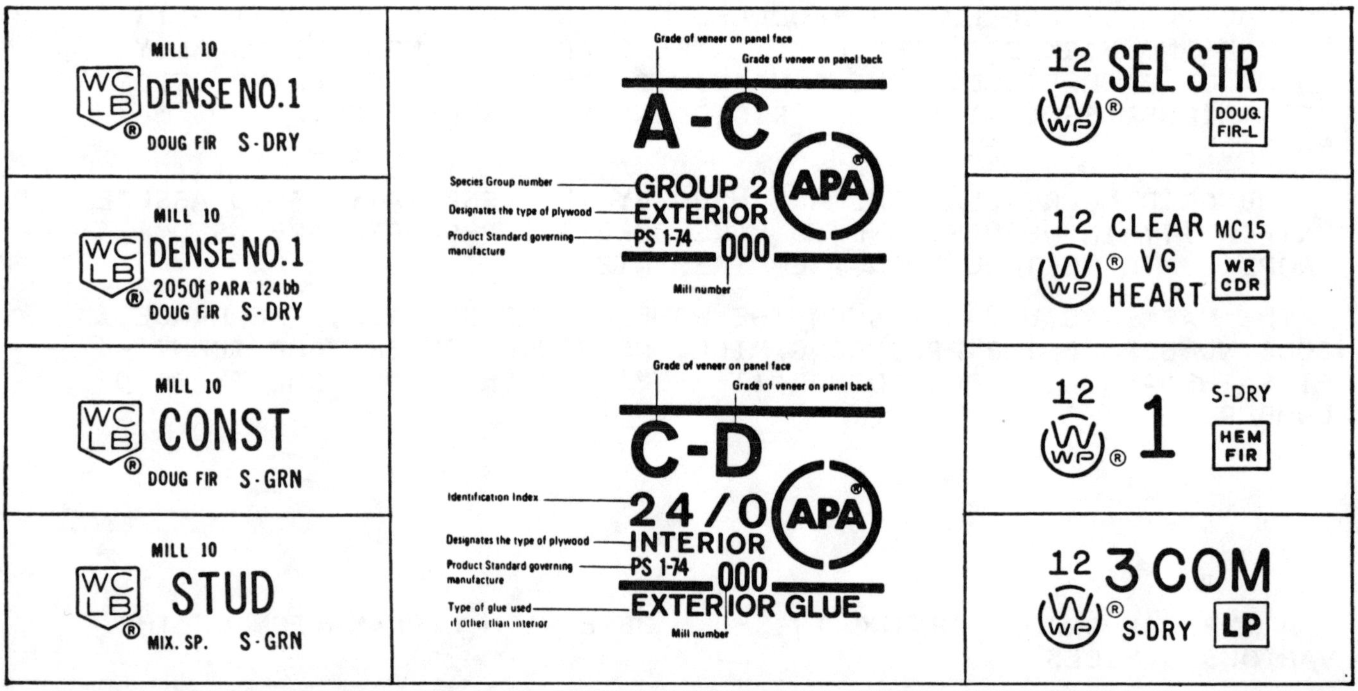

COURTESY __SUCCESSFUL HOW TO BUILD YOUR OWN HOME__, BY ROBERT C. RESCHKE (STRUCTURES PUBLISHING CO.)

SEE: *LUMBER GRADING, LUMBER KINDS, PLYWOOD, PLYWOOD TERMS.*

MANUFACTURER'S DIRECTIONS-----

MOST MANUFACTURERS OF BUILDING PRODUCTS ISSUE OR INCLUDE COMPLETE INSTRUCTIONS FOR THEIR USE. IT IS WISE TO ALWAYS FOLLOW THE DIRECTIONS AND NEVER TRY TO SUBSTITUTE OR USE A PRODUCT FOR A PURPOSE FOR WHICH IT IS NOT DESIGNED OR INTENDED.

UNLESS YOU FOLLOW DIRECTIONS AND SUGGESTIONS, WARRANTIES AND GUARANTEES FOR PRODUCTS AND MATERIALS ARE NOT VALID.

LUMBER SIZES-----

WHEN YOU DISCUSS LUMBER WITH A CONTRACTOR, DEALER OR CRAFTSMAN THEY WILL SPEAK OF CERTAIN SIZES AND TYPES AS "TWO BY FOURS, ONE BY EIGHTS," ETC. THIS IS THE COMMON NAME OR DESIGNATION OF VARIOUS SIZED PIECES. FOR EXAMPLE, A 2 X 4 WOULD BE 2" THICK AND 4" WIDE. THIS IS CALLED THE NOMINAL SIZE WHICH IS CUT AT THE MILL. AS A RESULT OF SHRINKAGE, SEASONING, DRESSING OR PLANING ON ALL FOUR SIDES, THE FINISHED SIZE IS CUT DOWN. THE SKETCH BELOW SHOWS THE NOMINAL SIZES OF THE MOST COMMONLY USED TYPES AND THE ACCOMPANYING TABLE GIVES THE ACTUAL MINIMUM SIZE. (THIS FINISHED SIZE MAY VARY IN DIFFERENT AREAS.)

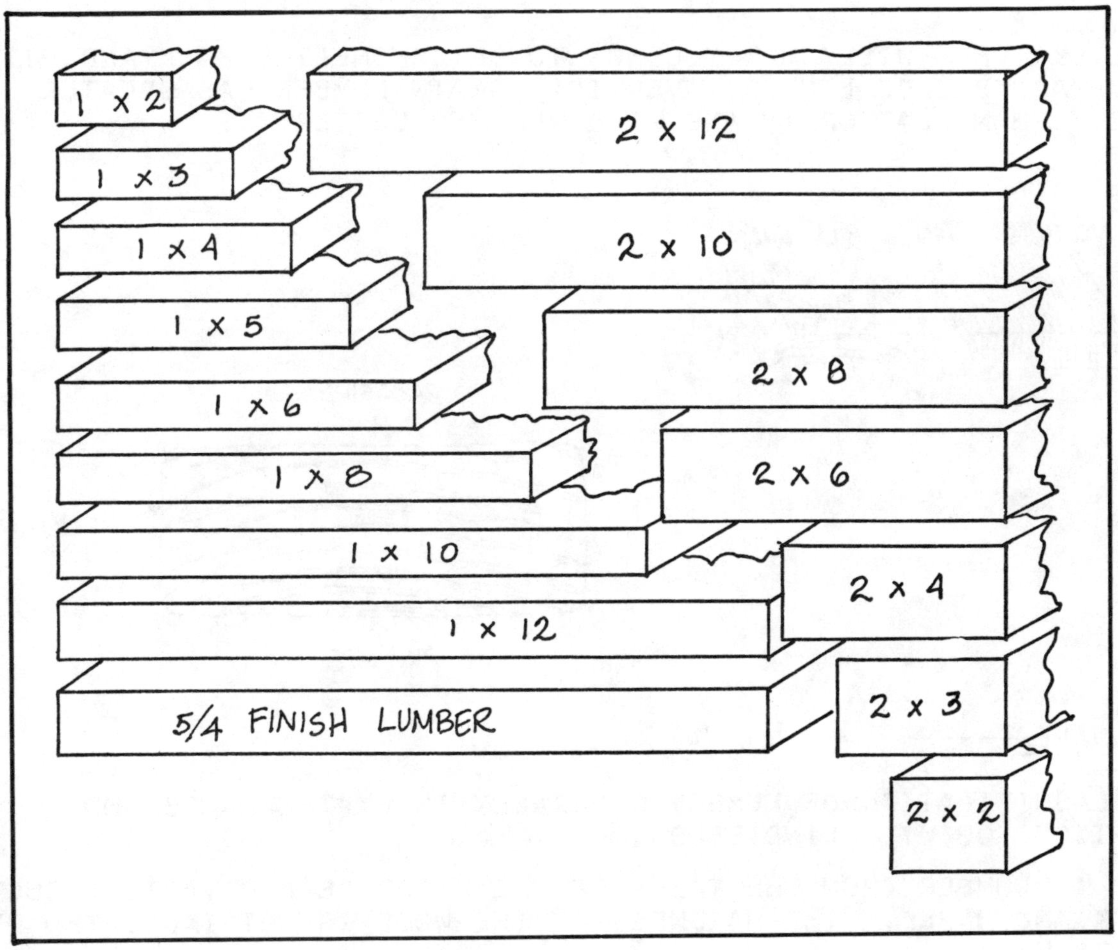

FINISH LUMBER IS DESIGNATED AS FIVE QUARTER OR LARGER THICKNESSES. THE 5/4 IS 1-1/16" THICK. 1" IS 3/4" THICK, 2" IS 1-1/2" THICK.

NOTE: "ACCOMPANYING TABLE" MENTIONED ABOVE IS ON NEXT PAGE.

LUMBER SIZES (continued)

NOMINAL SIZE	INCHES	MINIMUM CUT SIZE	INCHES
1 X 2	3/4 X 1-1/2	2 X 2	1-1/2 X 1-1/2
1 X 3	3/4 X 2-1/2	2 X 3	1-1/2 X 2-1/2
1 X 4	3/4 X 3-1/2	2 X 4	1-1/2 X 3-1/2
1 X 5	3/4 X 4-1/2	2 X 6	1-1/2 X 5-1/2
1 X 6	3/4 X 5-1/2	2 X 8	1-1/2 X 7-1/2
1 X 8	3/4 X 7-1/4	2 X 10	1-1/2 X 9-1/4
1 X 10	3/4 X 9-1/4	2 X 12	1-1/2 X11-1/4
1 X 12	3/4 X11-1/4	5/4 X 12	1-1/16X11-1/4

LUMBER GRAIN-----

THESE SKETCHES WILL SHOW YOU HOW TO DETERMINE THE BARK AND PITH SIDES OF A PIECE OF LUMBER. VERTICAL GRAIN LUMBER HAS GREATER STABILITY THAN FLAT GRAIN AND WILL WEATHER BETTER.

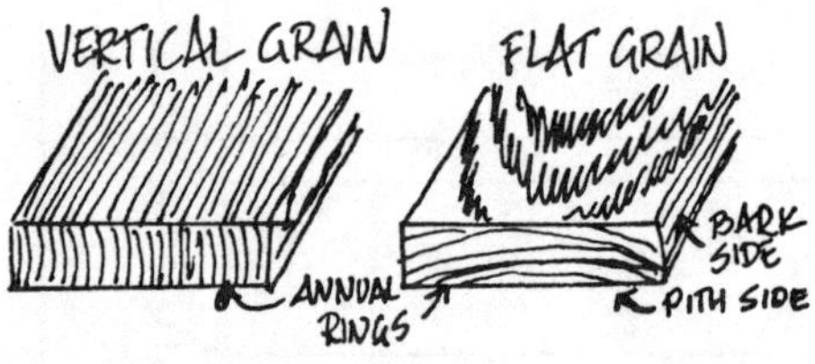

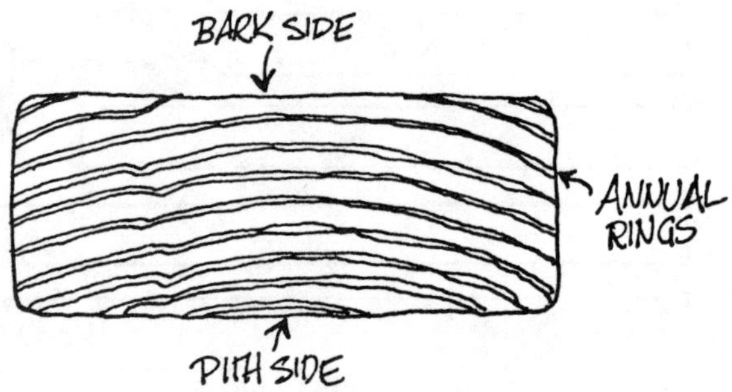

LUMBER NAMES-----

THE CLASSIFICATION OF LUMBER GENERALLY IS BY THE NAMES AND DESCRIPTION YOU WILL FIND BELOW.

HARDWOOD: LUMBER FROM THE TREES OF THE BROAD LEAFED VARIETY SUCH AS OAK AND MAPLE. THE HARDNESS OF THE WOOD IS NOT INDICATED BY THIS TERM.

SOFTWOOD: LUMBER MADE FROM THE TREES OF THE CONIFER TYPE KNOWN AS THE NEEDLE-BEARING TREES.

SAPWOOD: LUMBER MADE FROM OR CONTAINING PARTS OF OUTER LAYERS OF WOOD SURROUNDING THE TREE. LIGHT COLORED AS COMPARED TO THE WOOD AT THE CENTER OF THE TREE.

HEARTWOOD: LUMBER MADE FROM OR CONTAINING PARTS OF THE CENTER OR CORE OF THE TREE. GENERALLY DARK.

SEE: *LUMBER MARKING.*

LUMBER GRADING-----

LUMBER IS GRADED USING TERMS ACCORDING TO SIZES. GENERAL NAMES ARE SHOWN IN THIS LIST.

STRIPS: LUMBER LESS THAN TWO INCHES THICK AND SIX INCHES OR LESS IN WIDTH.

BOARDS: LUMBER LESS THAN TWO INCHES THICK BUT EIGHT INCHES OR MORE IN WIDTH.

(IN UTILIZING BOARDS AND STRIPS, APPEARANCE AND SERVICEABILITY ARE OF PRIMARY IMPORTANCE.)

DIMENSION: LUMBER TWO INCHES OR MORE, BUT LESS THAN FIVE INCHES, IN THICKNESS, OF ANY WIDTH.

JOISTS: DIMENSION OF SIZES ORDINARILY USED FOR FLOOR AND CEILING JOISTS.

TIMBERS: LUMBER AT LEAST FIVE INCHES OR MORE IN ANY DIMENSION.

(IN DIMENSION, JOISTS AND TIMBERS, THE STRENGTH, STIFFNESS, AND UNIFORMITY OF SIZE ARE OF PRIMARY IMPORTANCE; APPEARANCE IS SECONDARY. KNOTS, CHECKS, SPLITS AND SHAKE ARE PERMITTED IN THE VARIOUS GRADES DEPENDING ON THEIR EFFECT ON THE STRENGTH OF THE PIECE.)

LUMBER, SOFTWOOD PANELING-----

SOLID WOOD SOFTWOOD PANELINGS WERE ONCE USED EXTENSIVELY IN HOMES, PARTICULARLY IN RECREATION ROOMS, INFORMAL LIVING AREAS, KITCHENS AND BASEMENTS.

BUT THESE PANELING MATERIALS HAVE GIVEN WAY TO PLYWOOD AND HARDBOARD SHEET PANELINGS WHICH NORMALLY COME WITH A WOODGRAIN OR OTHER PATTERN PREFINISH. THE COMMON SHEET SIZE IS 4' BY 8'.

SOLID WOOD PANELINGS ONCE WERE FURNISHED IN A RANGE OF SIZES FROM 1X4 TO 1X12 WITH AN EXTENSIVE CHOICE OF FACE STYLES. EDGES WERE DRESSED AND MATCHED IN WHAT WERE NORMALLY KNOWN AS TONGUE-AND-GROOVE OR SHIPLAP JOINTS. TODAY, MANY BUILDING OR LUMBER SUPPLY BUSINESSES DO NOT STOCK SOLID WOOD PANELINGS AT ALL, AND THOSE THAT DO, USUALLY CARRY ONLY LIMITED SIZES IN PINE, CEDAR OR REDWOOD.

IN CONTRAST, THERE IS ALMOST AN UNLIMITED CHOICE OF PANELING STYLES IN PLYWOOD AND HARDBOARD SHEETS. MANY OF THESE PRE-FINISHED STYLES AND COLORS HAVE WOODGRAINS THAT SIMULATE THOSE OF REAL WOOD SPECIES. BUT IN ADDITION, IT IS ALSO COMMON PRACTICE AMONG SHEET PANELING MANUFACTURERS TO DEVELOP PRE-FINISHES FOR PANELS WHICH ARE NAMED AS THOUGH THEY WERE WOOD SPECIES BUT WHICH ACTUALLY ARE CREATED NAMES NOT BASED ON ANY ACTUAL SPECIES OF WOOD.

LUMBER-----

IF YOU WANT TO UTILIZE THE LUMBER OR WOOD TYPE BEST SUITED FOR ANY PARTICULAR PURPOSE, THIS DESCRIPTIVE LIST WILL ENABLE YOU TO MAKE A SELECTION BY COMPARING THE VARIOUS KINDS AND TYPES THAT MAY BE AVAILABLE IN YOUR AREA.

- ASH - MEDIUM GRADE HARDWOOD. FOR INTERIOR FINISH.
- BASSWOOD - LIGHTWOOD, SOFTWOOD. FOR MILLWORK, SIDING AND CABINET CONSTRUCTION.
- BEECH - A HARDWOOD FOR FLOORING AND INTERIOR FINISH.
- CEDAR (WHITE) - SOFTWOOD FOR SHINGLES, SIDING, POSTS.
- CEDAR (RED) - SOFTWOOD FOR SHINGLES, SIDING, LIGHT FRAMING— ALSO USED FOR MOTHPROOF CLOSET LINING (INCENSE CEDAR).
- CHERRY - MEDIUM SOFTWOOD FOR PANELING, TRIM AND FLOORING.
- CYPRESS - HARD SOFTWOOD FOR PANELING AND MISCELLANEOUS USES.
- FIR - SOFTWOOD FOR SIDING, FRAMING EXTERIOR TRIM, GUTTER, FLOORING.
- GUM - MEDIUM HARDWOOD FOR INTERIOR TRIM, MILLWORK. ALSO AS IMITATION WALNUT AND MAHOGANY.
- HICKORY - NOT USED IN CONSTRUCTION. FOR TOOL HANDLES, ETC.
- LARCH - SOFTWOOD FOR FRAMING AND MILLWORK.
- MAPLE - HARDWOOD FOR FLOORING AND FINISH.
- OAK - HARDWOOD FOR FLOORING, FINISH AND CABINET WORK.
- PINE (WHITE) - SOFTWOOD FOR MILLWORK, SHEETING, SUBFLOORING. COMMON GRADES USED FOR FRAMING, BOARDS, ETC.
- PINE (PONDEROSA) - SOFTWOOD USED FOR MILLWORK AND TRIM.
- PINE (SOUTHERN YELLOW) - SOFTWOOD USED FOR ALL TYPES OF LUMBER IN HOME BUILDING.
- PINE (LODGEPOLE) - SOFTWOOD FOR SIDING, FRAMING, BOARDS AND TIMBERS.
- POPLAR - SOFTWOOD FOR TRIM, MILLWORK AND CABINETS.
- REDWOOD - SOFTWOOD FOR SIDING, TRIM, ALL OUTSIDE WORK.
- SPRUCE (SITKA) - HARDWOOD FOR SIDING, MILLWORK, FINISH AND GENERAL HOME BUILDING.
- SPRUCE (ENGLEMANN) - HARDWOOD SIDING, GENERAL CONSTRUCTION, MILLWORK.

SEE: *WOOD*.
ALSO: SUCCESSFUL WOOD BOOK (STRUCTURES PUBLISHING COMPANY)

LUMBER ABBREVIATIONS----

HERE ARE SOME OF THE COMMON ABBREVIATIONS YOU WILL FIND WHEN YOU TALK BUILDING AND LUMBER WITH TRADES PEOPLE.

AD	AIR-DRIED
BD.	BOARD
BD. FT.	BOARD FEET
BEV.	BEVELLED
B.M.	BOARD (FOOT) MEASURE
BTR.	BETTER
CLG.	CEILING
CLR OR CL	CLEAR
COM.	COMMON
C.M.	CENTER MATCHED
CSG.	CASING
DIM.	DIMENSION
EM	END MATCHED
F.G.	FLAT GRAIN
FLG.	FLOORING
FT. B.M.	FEET BOARD MEASURE
FT. S.M.	FEET SURFACE MEASURE
HDWD.	HARDWOOD
HRT.	HEART
KD	KILN-DRIED
LBR.	LUMBER
LGTH.	LENGTH
M B.M.	THOUSAND (FEET) BOARD MEASURE
M.L.	MIXED LENGTHS
MLDG.	MOLDING
M.W.	MIXED WIDTHS
PAT.	PATTERN
R.W.	RANDOM WIDTHS
RDM.	RANDOM
RFG.	ROOFING
R.L. OR R/L	RANDOM LENGTHS
SDG. OR SIDG. OR S/G	SIDING
S.E.	SQUARE-EDGE
SEL.	SELECT
S.F.	SURFACE FOOT (1 SQUARE FOOT)
SFTWD.	SOFTWOOD
SND.	SOUND
STND.	STAINED
STR.	STRUCTURAL
T&G	TONGUED AND GROOVED
TBRS.	TIMBERS

LUMBER PRICING-----

BEFORE YOU GET VERY FAR ALONG IN BUILDING, REMODELING OR RENOVATING A HOME, YOU WILL BE DISCUSSING LUMBER PRICES WITH THE CONTRACTOR, DEALER OR CARPENTERS.

THE COMMON UNIT OF MEASURE IS A BOARD FOOT. ONE BOARD FOOT IS A PIECE OF LUMBER 12 INCHES WIDE, 1 INCH THICK AND 1 FOOT LONG. THUS A PIECE OF LUMBER 1" THICK, 12" WIDE AND 12' LONG WOULD CONTAIN 12 SQUARE FEET OR 12 BOARD FEET OF LUMBER.

WHILE A COMMON BOARD AS ABOVE IS CALCULATED AND SOLD AS 1"X12"X12' AND CONTAINS 12 SQUARE OR BOARD FEET AND PRICED ON THAT BASIS, THE ACTUAL THICKNESS AND WIDTH ARE LESS. THE LENGTHS ARE ALWAYS CONSTANT AS INDICATED.

USUALLY ON BUILDING PROJECTS THE PRICE IS FIGURED AND QUOTED ON THE BASIS OF 1,000 BOARD FEET. PRESENTLY, HOWEVER, MODERN MARKETING CENTERS ARE QUOTING PRICES ON PIECES. PER PIECE PRICES ARE LIKELY TO BE HIGHER THAN BOARD FEET PRICES BECAUSE PIECES USUALLY ARE SOLD IN SMALL LOTS.

LAND CONTRACTS-----

LAND CONTRACTS ARE OFTEN USED IN PURCHASING PROPERTY WHEN THE BUYER HAS LIMITED FUNDS FOR A DOWN PAYMENT AND THE SELLER IS IN POSITION TO FINANCE THE SALE. INTEREST AND PAYMENTS MAY BE DICTATED BY THE SELLER WITH PAYMENTS MADE TO THE SELLER OR HIS DESIGNATED AGENT. THE DEED AND TITLE IS HELD BY THE ORIGINAL OWNER UNTIL ALL THE PAYMENTS ARE MADE AND THE PROPERTY IS PAID IN FULL. VARIOUS ARRANGEMENTS ARE MADE IN DEALS OF THIS TYPE WITH REGARD TO TAX PAYMENTS, ETC.

LIGHTING-----

THIS CHART WILL ASSIST YOU IN SELECTING THE PROPER LIGHTING LEVELS FOR VARIOUS SIZED ROOMS IN YOUR HOME. THREE TYPES OF LIGHTING ARE INDICATED WITH LAMP WATTAGE SUGGESTIONS. THE LAMP SIZES AND WATTAGES MAY BE REDUCED IN THE INTEREST OF ENERGY CONSERVATION.

	*A. Surface and Pendent Fixtures**		*B. Recessed Fixtures**	*C. Wall Lighting***
Room Size	*Minimum Size of Shield*	*Minimum Bulb(s) or Watts per sq. ft.*	*Minimum Wattage or Watts per sq. ft.*	*Length of light source; or Wattage and spacing of fixtures*
VERY SMALL (Up to 125 sq. ft.)	12" to 15"	One 100-watt or Three 40-watt	150-watt fixture for each 75 sq. ft. floor area or fraction thereof	6' or 75-watt fixtures on 3' centers
AVERAGE (125 to 225 sq. ft.)	15" to 17"	One 150-watt or Four 40-watt	150-watt fixture for each 75 sq. ft. floor area or fraction thereof	8' to 12' or 75-watt fixtures on 3' centers
LARGE (Over 225 sq. ft.)	17" to 22"	100 watts for each 100 sq. ft. *footnote	150-watt fixture for each 75 sq. ft. floor area or fraction thereof	16' to 20' or 75-watt fixtures on 3' centers

*More than one fixture is usually needed on ceilings and/or walls for minimum general lighting. Dining area fixture should provide predominant downlight on table.

**Fixture recommendation applies to lighting one long wall. For most effective design, consult electric utility, lighting fixture distributor, or manufacturers' booklets.

COURTESY OF AMERICAN HOME LIGHTING INSTITUTE.

SEE: *LIGHTING TERMS, LIGHT REFLECTIVITY.*

LIGHTING TERMS-----

WHEN YOU CONSIDER LIGHTING FOR YOUR HOME, YOU WILL FIND MANY OF THESE TERMS.

- BALLAST - A CURRENT-REGULATING AUXILIARY DEVICE REQUIRED TO OPERATE A FLUORESCENT LAMP.
- BRIGHTNESS - THE DEGREE OF APPARENT LIGHTNESS OF ANY SURFACE EMITTING OR REFLECTING LIGHT.
- CANDLEPOWER - LUMINOUS INTENSITY—STRENGTH OF LIGHT.
- COFFER - RECESSED PANEL IN CEILING, OR DOME.
- COOL WHITE - FLUORESCENT LAMP "WHITER" THAN FILAMENT LAMP; BLENDS WELL WITH DAYLIGHT.
- DIFFUSION - THE SCATTERING OF LIGHT RAYS, SO THAT THE LIGHT IS EMITTED OR REFLECTED FROM ALL DIRECTIONS.
- DIRECT LIGHTING - A LIGHTING SYSTEM IN WHICH ALL OR NEARLY ALL OF THE LIGHT IS DISTRIBUTED DOWNWARD.
  (FIXTURES THAT SEND SOMEWHAT MORE LIGHT UPWARD BUT ARE STILL THE DIRECT TYPE ARE CALLED SEMI-DIRECT.)
- DOWNLIGHT - SMALL, CONCENTRATED DIRECT LIGHTING UNIT RECESSED IN CEILING.
- FLUORESCENT LAMP - A LIGHT SOURCE IN WHICH LIGHT IS PRODUCED BY A FLUORESCENT POWDER (PHOSPHOR) COATED ON THE INNER SURFACE OF A GLASS TUBE.
- FOOTCANDLE (FT-C) - IT IS THE APPROXIMATE ILLUMINATION PRODUCED BY A PLUMBER'S CANDLE AT A DISTANCE OF ONE FOOT.
- GLARE - LIGHT OR BRIGHTNESS IN THE FIELD OF VIEW WHICH CAUSES VISUAL DISCOMFORT AND/OR INTERFERES WITH SEEING.
- LUMEN - A QUANTITIVE UNIT FOR MEASURING LIGHT OUTPUT.
- LUMILINE - A TUBULAR INCANDESCENT LIGHT SOURCE WHICH HAS A FILAMENT EXTENDING THE LENGTH OF THE TUBE AND CONNECTS AT EACH END TO A DISC BASE.
- LUMINAIRE - A COMPLETE LIGHTING UNIT: CEILING FIXTURE, WALL BRACKET, PORTABLE LAMP, OR BUILT-IN OR APPLIED UNITS.
- SHIELDING - AN ARRANGEMENT OF LIGHT-CONTROLLING MATERIAL TO PREVENT THE DIRECT VIEW OF THE LIGHT SOURCE.
- VALANCE - A HORIZONTAL MEMBER MOUNTED ACROSS THE TOP OF A WINDOW TO CONCEAL LIGHT SOURCES.
- WARM WHITE - FLUORESCENT LAMP RESEMBLING THE COLOR OF FILAMENT LAMP WHICH BLENDS WELL WITH IT.
- WATT - UNIT OF MEASURE OF ELECTRICAL POWER—THE RATE OF WORK DONE.

LOCK TYPES-----

YOUR HOME SECURITY STARTS WITH PROPER LOCKS ON ALL EXTERIOR DOORS. THE FOLLOWING DESCRIPTIVE NAMES COVERING THE VARIOUS TYPES WILL HELP YOU IN SELECTING THE PROPER LOCKS.

KNOB LATCH TYPE—RECOMMENDED FOR PASSAGE AND CLOSED DOORS. OUTSIDE AND INSIDE KNOBS ARE FREE. LATCH BOLT RETRACTS WHEN EITHER KNOB IS TURNED.

BUTTON TYPE—FOR BASEMENT DOORS. INSIDE BUTTON LOCKS OUTSIDE KNOB. TURNING INSIDE KNOB AND CLOSING DOOR RELEASES BUTTON.

PRIVACY LOCK—BATHROOM. PUSHING INSIDE BUTTON LOCKS OUTSIDE KNOB. TURNING INSIDE KNOB AND CLOSING DOOR RELEASES THE BUTTON, PREVENTS LOCKOUT.

KEY LOCK TYPE—FOR FRONT DOORS. TURNING BUTTON ON THE INSIDE KNOB DOES NOT RELEASE UNLESS TURNED TO VERTICAL POSITION. KEY OPENS LOCK FROM OUTSIDE.

PUSH BUTTON KEY LOCK—FOR ENTRANCE DOORS. BUTTON IN THE INSIDE KNOB LOCKS OUTSIDE KNOB. TURNING INSIDE KNOB OR KEY RELEASES BUTTON.

KEY CONTROL LOCK—RECOMMENDED FOR UTILITY DOORS. MAY BE ACTIVATED BY KEY IN OUTSIDE KNOB. INSIDE KNOB IS FREE FOR EXIT. LATCH BOLT DEADLOCKS WHEN DOOR IS CLOSED.

LOCK BOLTS-----

SPRINGLATCH—BEVELED, SPRING OPERATED BOLT PUSHED BACK BY PRESSURE OR RETRACTED BY KNOB OR HANDLE.

DEADLATCH—TYPE WITH SMALL LATCH WHICH DEADLOCKS THE LATCHBOLT.

DEADBOLT—MANUALLY OPERATED BOLT WHICH IS LOCKED BY TURNING A KNOB INSIDE THE HOUSE, OR BY TURNING A KEY OUTSIDE.

KEY PLANS-----

ALL ALIKE—ANY KEY OPENING ALL LOCKS. TUMBLERS ARE IDENTICAL IN EACH LOCK OR DOOR.

ALL DIFFERENT—EACH KEY OPENS ONLY ONE LOCK. USUALLY STOCK LOCKS WORK LIKE THIS.

MASTER KEY—ALL LOCKS DIFFERENT, WITH A MASTER KEY WHICH WILL OPEN ALL LOCKS. USUALLY MADE TO ORDER ONLY.

LOCKS-----

THE CUT OUT PATTERNS MAY DETERMINE THE TYPE YOU REQUIRE FOR CERTAIN DOORS.

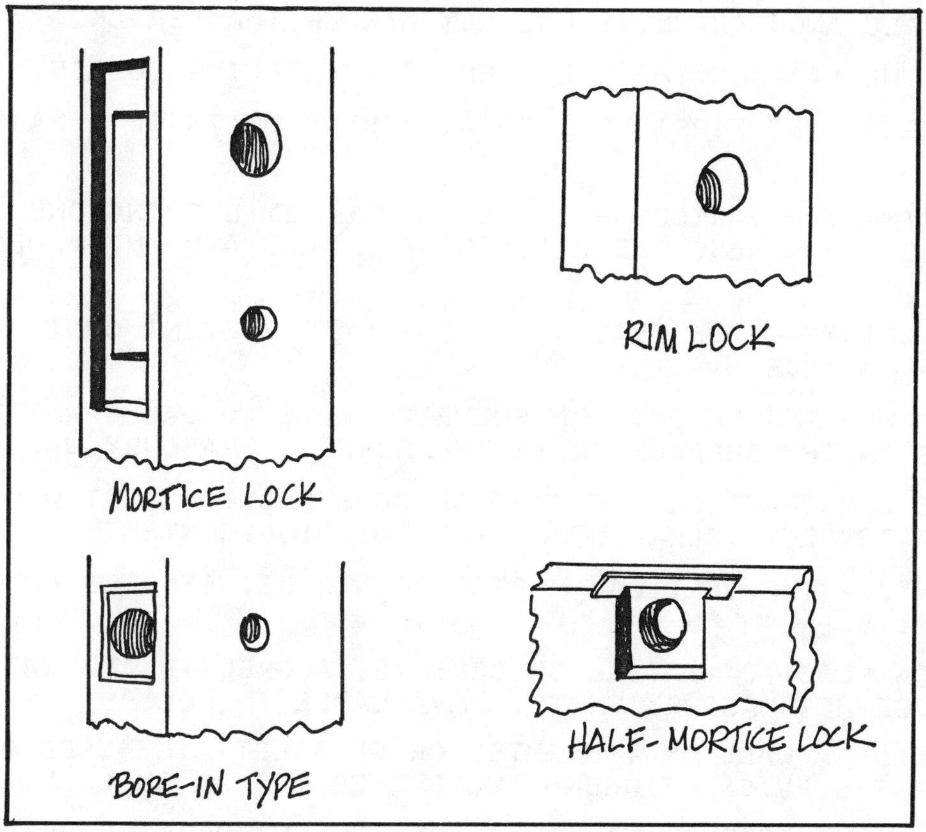

METAL LATH, USES-----

THE SKETCH BELOW SHOWS THE VARIOUS USES OF METAL LATH ACCESSORIES FOR USE IN PLASTERED WALL CONSTRUCTION.

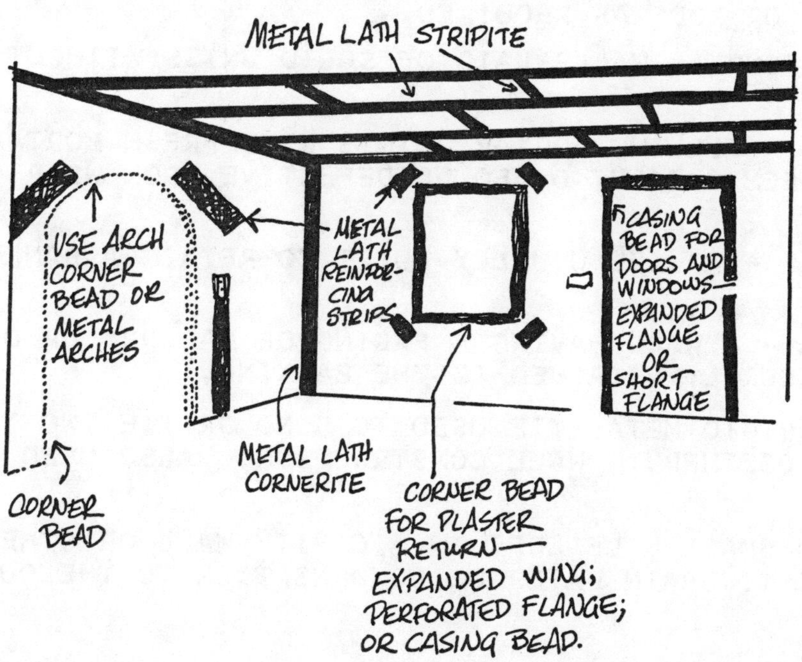

MASONRY TERMS-----

THESE ARE SOME OF THE COMMONLY USED TERMS PERTAINING TO MASONRY CONSTRUCTION.

- ASHLAR - RECTANGULAR UNITS USUALLY LARGER THAN BRICK.
- BACKUP - THE PART OF THE WALL BEHIND THE FACING COURSE.
- BASE COURSE - THE FIRST HORIZONTAL ROW OF UNITS OF A MASONRY WALL.
- BOND - TYING THE VARIOUS PARTS OF A WALL BY LAPPING ONE UNIT OVER ANOTHER. ALSO THE PATTERN OF A WALL AND THE ADHESION OF MORTAR.
- BREAKING JOINTS - THE PLACING OF UNITS TO PREVENT CONTINUOUS VERTICAL JOINTS.
- EFFLORESCENCE - A DEPOSIT OF SOLUBLE SALTS, USUALLY WHITE, THAT APPEARS ON THE SURFACE OR IN THE PORES OF MASONRY WALLS.
- FIREPROOF CONSTRUCTION - A TYPE OF CONSTRUCTION THAT WILL WITHSTAND A COMPLETE BURN-OUT OF THE BUILDING CONTENTS.
- HEAD JOINTS - THE VERTICAL MORTAR JOINTS BETWEEN THE ENDS OF WALL UNITS.
- MORTAR - A PLASTIC MIXTURE OF CEMENTS, AGGREGATE AND WATER TO BIND TOGETHER MASONRY UNITS. AVAILABLE IN COLORS.
- PARTY WALL - A WALL ON THE INTERIOR OF A LOT LINE USED OR ADAPTED FOR JOINT SERVICE BETWEEN TWO PROPERTIES OR BUILDINGS.
- PILASTER - A THICKENED WALL SECTION, COLUMN OR PIER BUILT AS PART OF A WALL.
- POINTING - TROWELING MORTAR INTO JOINTS AFTER WALL IS LAID.
- REINFORCED CONCRETE MASONRY - MASONRY WHICH IS REINFORCED WITH STEEL WIRES OR RODS AS REQUIRED.
- SOLID UNIT MASONRY - WALLS LAID OF SOLID UNITS WITHOUT REINFORCING..
- TUCK POINTING - FILLING MORTAR JOINTS WITH FRESH MORTAR AFTER THEY HAVE BEEN CUT OUT OR FOUND DEFECTIVE FROM WEAR OR EROSION.
- RETAINING WALL - A WALL USUALLY BUILT TO RETAIN A BANK OF EARTH OR FILL.
- VENEERED WALL - A WALL HAVING A FACING OF MASONRY OR OTHER MATERIAL SECURELY ATTACHED TO THE BACKING.
- WALL TIE - A RIGID METAL TIE USED TO BOND OR TIE TWO TYPES OF MATERIALS TOGETHER IN WALL CONSTRUCTION. ALSO USED IN CAVITY WALLS.
- WEEP HOLE - A SMALL HOLE LEFT IN A CAVITY WALL OR OTHER TYPES IF REQUIRED TO DRAIN WATER OR CONDENSATION TO THE OUTSIDE.

MATERIAL LISTS-----

IT REQUIRES CONSIDERABLE TIME AND EXPERTISE TO COMPILE A LIST OF ALL THE MATERIALS NEEDED TO BUILD THE AVERAGE HOUSE. WHEN YOUR BUILDER PREPARES YOUR BID HE USUALLY ARRIVES AT HIS PRICE BY LISTING ALL THE MATERIALS REQUIRED AND USES THE CURRENT COSTS PLUS THE LABOR REQUIRED TO GET THE TOTAL COST FOR YOU. FOR QUICK ESTIMATES, SOME BUILDERS MAY USE THE CUBIC FOOT OR SQUARE FOOT PRICES ON CONSTRUCTION OF THE SAME GENERAL TYPE.

WHEN YOU PURCHASE A SET OF READY MADE HOUSE PLANS FROM ONE OF THE LEADING HOUSE PLAN CONCERNS, YOU WILL USUALLY RECEIVE COMPLETE MATERIAL LISTS WITH THE PLANS.

MECHANICS LIENS-----

WHEN MATERIAL OR LABOR USED ON YOUR JOB IS NOT PAID FOR, THE SUPPLIER OR LABORER MAY PLACE A LIEN AGAINST THE PROPERTY AND CAN LEGALLY COLLECT THE AMOUNT DUE HIM.

MORTGAGES, GRADUATED-----

THIS TYPE MORTGAGE IS SOMETIMES DESIGNATED AS A VARIABLE RATE MORTGAGE. THEY ARE AVAILABLE IN MANY AREAS FROM SOME LENDING AGENCIES. THIS TYPE OF FINANCING MAY BE HELPFUL TO MANY INDIVIDUALS DUE TO THE PRESENT HIGH PRICES OF REAL PROPERTY AND BUILDING. THE INTEREST RATES VARY IN THIS TYPE OF FINANCING, AND IN THE LONG RUN ARE SLIGHTLY HIGHER. MORTGAGES OF THIS TYPE ENABLE THE HIGHLY QUALIFIED PURCHASER TO BUY OR BUILD A HIGHER PRICED HOME THAN HE WOULD UNDER NORMAL TYPE FINANCING.

MORTGAGES, PLAIN LANGUAGE-----

PLAIN LANGUAGE MORTGAGES AND SIMPLIFIED DOCUMENTS COVERING REAL ESTATE TRANSACTIONS ARE NOW BEING ACTIVELY PROMOTED BY THE FEDERAL NATIONAL MORTGAGE ASSOCIATION. THIS TYPE MORTGAGE WILL ELIMINATE MUCH OF THE LEGAL TERMINOLOGY NOW USED.

MORTGAGE, OPEN END-----

AN OPEN END MORTGAGE IS A DOCUMENT PREPARED OR ALTERED TO PERMIT THE USE OF ADDITIONAL FUNDS UP TO THE LIMIT OF THE EQUITY OR APPRAISAL VALUE OF THE PROPERTY THAT IS INITIALLY COVERED. DEPENDING ON THE POLICY OF THE LENDING INSTITUTION, THESE FUNDS MAY BE

MORTGAGE, OPEN END (continued)

USED FOR PROPERTY IMPROVEMENT OR VARIOUS TYPES OF EMERGENCIES OR SITUATIONS REQUIRING FUNDS.

MORTGAGE, WRAP AROUND-----

A WRAP AROUND MORTGAGE IS SOMETIMES KNOWN AS AN ALL INCLUSIVE MORTGAGE. IN REALITY IT IS A SECOND MORTGAGE AS IT TENDS TO WRAP AROUND AN EXISTING MORTGAGE. USUALLY THE WRAP AROUND PORTION OF A TRANSACTION INCURS A HIGHER RATE OF INTEREST THAN THE ORIGINAL OR FIRST MORTGAGE.

SEE: *SECOND MORTGAGES*.

PROPERTY OWNERSHIP-----

THIS SCHEDULE OF OWNERSHIP USUALLY BECOMES EFFECTIVE WHEN AN OWNER IS DECEASED WITHOUT A WILL TO DETERMINE THE SUBSEQUENT OWNERSHIP OF THE PROPERTY.

DECEASED WITH SURVIVING SPOUSE AND CHILDREN:
- ONE CHILD WOULD GET HALF OF ESTATE, SPOUSE WOULD GET HALF OF ESTATE.
- TWO OR MORE CHILDREN WOULD GET TWO-THIRDS OF ESTATE, SPOUSE WOULD GET ONE-THIRD OF ESTATE.

WITH SURVIVING SPOUSE AND NO CHILDREN:
- PARENTS WOULD GET ONE-QUARTER OF ESTATE, SPOUSE WOULD GET THREE-QUARTERS OF ESTATE.

WITH SURVIVING SPOUSE AND NO PARENTS:
- SPOUSE WOULD GET ENTIRE ESTATE.

DECEASED WITH NO SURVIVING SPOUSE:
- CHILDREN WOULD GET ENTIRE ESTATE;
- WITH NO CHILDREN, PARENTS WOULD GET ENTIRE ESTATE;
- WITH NO PARENTS, BROTHERS AND SISTERS WOULD GET ENTIRE ESTATE;
- WITH NO BROTHERS AND SISTERS, GRANDPARENTS WOULD GET ENTIRE ESTATE;
- WITH NO GRANDPARENTS, LINEAL DESCENDANTS WOULD GET ENTIRE ESTATE;
- WITH NO LINEAL DESCENDANTS, NEXT OF KIN GETS ENTIRE ESTATE;
- WITH NO NEXT OF KIN, STEPCHILDREN WOULD GET ENTIRE ESTATE;
- WITH NO STEPCHILDREN, STATE GETS ENTIRE ESTATE.

NOTE: THIS IS ONLY A GUIDE. BE SURE TO CHECK YOUR LOCAL LAWS AS EACH STATE HAS DIFFERENT AND SPECIFIC ESTATE REGULATIONS AND SUCCESSION LAWS.

MOLDINGS, STANDARD PATTERNS-----

THE STANDARD PATTERNS OF INTERIOR WOOD MOLDINGS OR TRIM GENERALLY USED IN NEW HOMES, REMODELING OR RENOVATING OLDER HOMES ARE SHOWN IN THE NUMBERED SKETCHES BELOW.

THESE MOLDINGS ARE USUALLY AVAILABLE IN MOST AREAS IN EASY TO WORK SOFTWOODS IN STANDARD LENGTHS. THEY MAY ALSO BE FOUND IN VARIOUS LENGTHS OF FINGER JOINTED STOCK, THEN PAINTED SO THE JOINTS WILL NOT SHOW.

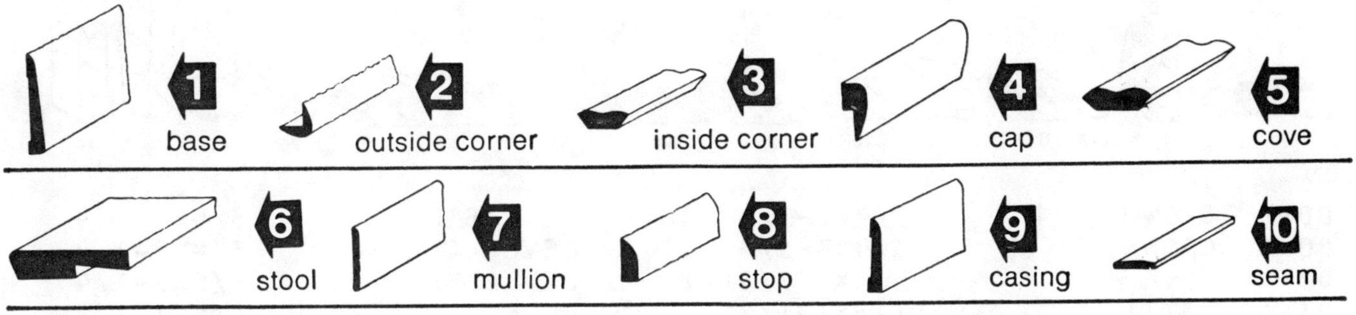

THE SKETCH BELOW IDENTIFIES THE MOLDINGS BY NUMBERS AND INDICATES WHERE THEY ARE USUALLY USED FOR INTERIOR TRIM AND DECORATION.

COURTESY FINDING & FIXING THE OLDER HOME (STRUCTURES PUBLISHING CO.)

SEE: *MOLDINGS, HARDBOARD MOLDINGS.*

MOLDINGS-----

THE FOLLOWING ILLUSTRATED MOLDINGS (WITH COMMON SIZES) ARE NORMALLY IN STOCK IN MOST AREAS. OTHER TYPES AND SIZES MAY BE MADE TO ORDER BY SOME PLANTS. A FEW OF THE COMMONLY USED TYPES ALSO MAY BE AVAILABLE IN PREFINISHED STAINS, COLORS OR PATTERNS.

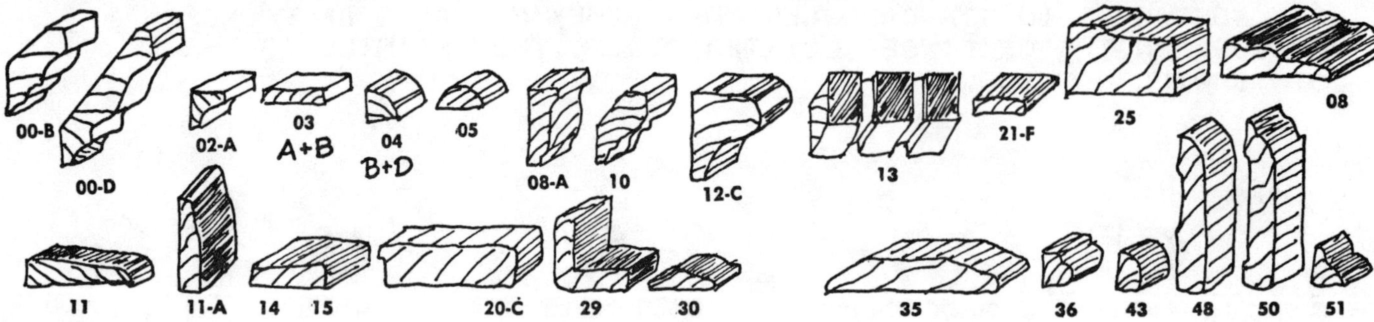

00-B	CROWN	3/4X2-3/4	08	OG CASING	3/4X2-1/4
00-D	CROWN	3/4X3-5/8	11	BEVEL CASING	3/4X2-1/4
02-A	COVE	3/4X 7/8	11-A	BEVEL BASE F.J.	5/8X2-1/4
03-A	LATTICE	5/16X1-1/8	14	ROUND EDGE STOP	1/2X1-1/8
03-B	LATTICE	5/16X1=3/8	15	RD. EDGE STOP	1/2X1-3/8
04-B	QUARTER ROUND	1/2X 1/2	20-C	CASEMENT STOOL	3/4X3-3/16
04-D	QUARTER ROUND	3/4X 3/4	29	CORNER MOLD	1-3/8X1-
05	HALF ROUND	5/16X 5/8	30		3/8X3/8
08-A	BARGE	3/4X1-3/4	35	PANEL MOLD	3/8X1-3/8
10	BED	3/4X1-3/4	36	THRESHOLDS	5/8X3-5/8-3'1"
12-C	BED	1-5/8X2	43	BASE MOLD	5/8 X 3/4
13	DENTIL MOLD	1-1/8X1-1/2	48	BASE SHOE	1/2 X 3/4
21-F	SCREEN MOLD	1/4X 3/4	50	RANCH BASE	5/8X3-1/4
25	OUTSIDE CASING	1-3/8X2	51	O.G. BASE	5/8X3-1/2
				BASE SHOE	3/4 X 3/4

Notes:

MOLDINGS, BASIC TYPES-----

THE MOLDINGS SHOWN BELOW ARE THE BASIC TYPES, WITH THE ARCHITECTURALLY CORRECT NAMES INDICATED.

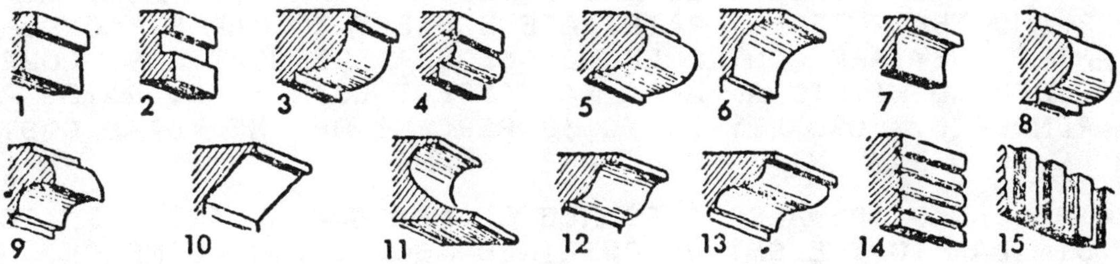

1. FILLET, FASCIA
2. SUNK FILLET
3. 1/4°, OVOLA
4. BEAD, ASTRAGAL
5. TORUS
6. CAVETTO
7. CONGE
8. OVOLO
9. BEAK
10. SPLAY
11. SCOTIA
12. CYMA RECTA
13. CYMA REVERSA
14. REED
15. DENTIL

TILE-----

PLASTIC WALL TILE IS GENERALLY USED IN BATHS AND KITCHENS, ALTHOUGH IT MAY BE USED ON ANY SMOOTH WALL. THIS TYPE OF TILE IS USED IN MANY REMODELING AND RENOVATION PROJECTS WHERE DAMAGED OR WORN-OUT WALLS NEED TO BE CONVERTED TO A MORE MODERN APPEARANCE. THE TWO BASIC SIZES AND FITTINGS AVAILABLE ARE ILLUSTRATED BELOW.

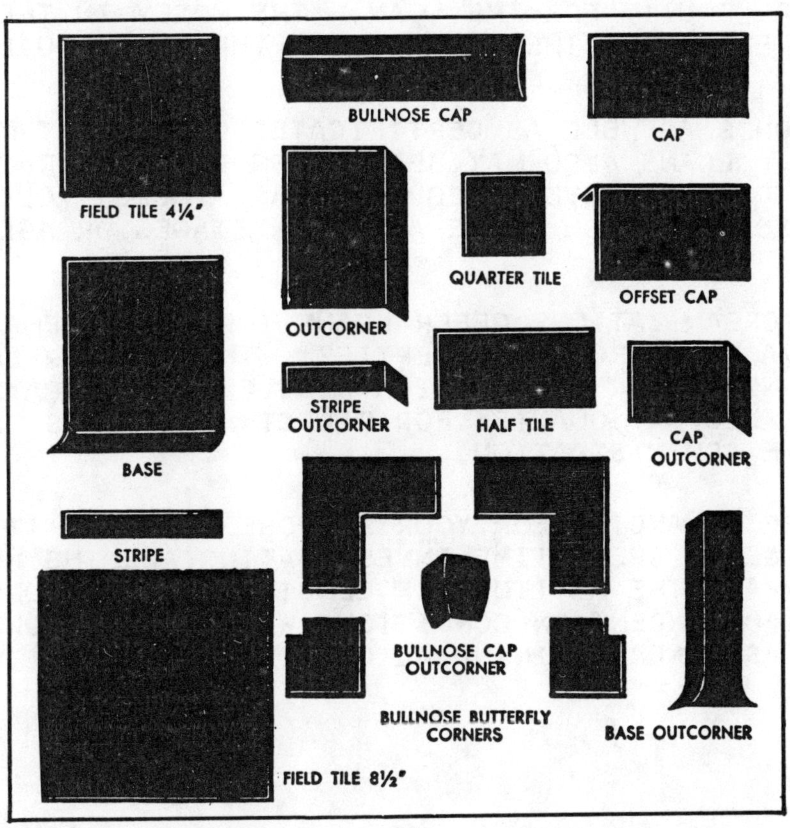

MONEY FOR BUYING, BUILDING OR REMODELING-----

MONEY FOR BUYING, BUILDING OR REMODELING MAY BE OBTAINED FROM SEVERAL SOURCES. SAVINGS AND LOAN ASSOCIATIONS DOMINATE THIS FIELD AS THEY ARE IN BUSINESS PRIMARILY TO FINANCE REAL ESTATE IN ONE FORM OR ANOTHER. POLICIES AND RATES MAY VARY FROM ONE AREA TO ANOTHER AND THE INTEREST RATES ARE SUBJECT TO MARKET CONDITIONS. WHILE THEY ARE CONTROLLED BY FEDERAL REGULATIONS, COMPETITION AND DEMAND WITHIN AN AREA MAY AFFECT LOAN FEES, RATES AND SERVICES RENDERED. USUALLY 10 TO 30 PERCENT OF THE TOTAL COST IS REQUIRED AS DOWN PAYMENT TO PURCHASE A HOME.

COMMERCIAL BANKS ARE ALSO A SOURCE OF MORTGAGE LOANS WITH TERMS SOMEWHAT SIMILAR TO THE SET-UP OUTLINED ABOVE. ON LOANS OF THIS TYPE THE TITLE IS TRANSFERRED TO THE NEW OWNER AND RECORDED IN HIS NAME. THE DEED, DEED COPY, OR ABSTRACT IS HELD BY THE LENDER. IN THE CASE OF NEW HOME LOANS, BOTH BANKS AND SAVINGS AND LOAN ASSOCIATIONS REQUIRE COMPLETE PLANS, SPECIFICATIONS AND ACCURATE COST ESTIMATES IN ORDER TO EVALUATE THE PROJECT AND DETERMINE LOAN VALUES. CONSTRUCTION LOANS REQUIRE RIGID AND PERIODICAL INSPECTIONS BY THE LENDING INSTITUTIONS IN ADDITION TO THE INSPECTION PROCEDURES REQUIRED BY BUILDING CODES AND REGULATIONS. LIFE INSURANCE COMPANIES AND MANY PRIVATE LOAN OFFICES ALSO MAKE LOANS ON REAL ESTATE AND ARE A GOOD SOURCE TO CONSIDER WHERE AVAILABLE. IN MANY CASES PRESENT LOANS MAY BE EXTENDED TO HANDLE FINANCING ON REMODELING PROJECTS. IN CASES OF THIS TYPE, A COMPLETE SKETCH, LAYOUT OR PLAN AS REQUIRED TO SECURE A BUILDING PERMIT, PLUS AN ACCURATE ESTIMATE, IS REQUIRED TO BE CONSIDERED BY THE LENDING ORGANIZATION.

PASSBOOK LOANS ARE MADE ON SAVINGS ACCOUNTS BY SURRENDERING YOUR PASSBOOK AS EQUITY FOR THE LOAN. THE MONEY IN THE ACCOUNT STILL BEARS INTEREST AND THE INTEREST ON THE LOAN USUALLY IS VERY LOW.

STOCKS AND BONDS AS WELL AS CERTIFICATES OF DEPOSIT MAY ALSO BE USED AS EQUITY. LOANS ALSO MAY BE MADE ON LIFE INSURANCE POLICIES THAT ARE OLD ENOUGH TO DEVELOP EQUITY. AS A RULE, YOUR POLICY WILL SHOW THE LOAN OR CASH VALUE AT GIVEN STAGES OR AGES OF THE POLICY.

SOME LENDING ORGANIZATIONS OFFER LOANS INSURED BY FHA OR GUARANTEED BY VA ON APPROVED PROPERTIES. THESE OFFER LOWER DOWN PAYMENTS AND LONGER TERMS WHICH REDUCE MONTHLY PAYMENTS. SOME BORROWERS ALSO MAY QUALIFY FOR DIRECT HOME LOANS FROM THE FARMERS HOME ADMINISTRATION.

IF YOU REQUIRE FINANCING FOR YOUR PROPOSED PROJECT IT WILL BE TO YOUR ADVANTAGE TO SPEND TIME INVESTIGATING ALL THE POSSIBILITIES IN YOUR AREA. THE ATTITUDE OF LENDING INSTITUTIONS MAY VARY FROM DAY TO DAY DEPENDENT ON CONDITIONS WITHIN THEIR OWN ORGANIZATION AS WELL AS GENERAL FINANCIAL CONDITIONS.

MEASUREMENTS U.S., LINEAR, SQUARE, CUBIC, LIQUID-----

LINEAR	
12 INCHES	1 FOOT
3 FEET	1 YARD
1760 YARDS	1 MILE
5280 FEET	1 MILE
16-1/2 FEET	1 ROD
5-1/2 YARDS	1 ROD
40 RODS	1 FURLONG

SQUARE	
144 SQUARE INCHES	1 SQUARE FOOT
9 SQUARE FEET	1 SQUARE YARD
30-1/4 SQUARE YARDS	1 SQUARE ROD
40 SQUARE RODS	1 ROOD
160 SQUARE RODS	1 ACRE
4 ROODS	1 ACRE
640 ACRES	1 SQUARE MILE

A "SQUARE" IS 100 SQUARE FEET.
TO GET THE AREA MULTIPLY THE WIDTH BY THE LENGTH.

CUBIC	CUBIC INCHES	CUBIC FEET
1 CUBIC	1	
1 CUBIC FOOT	1728	1
1 CUBIC YARD	45656	27

231 CUBIC INCHES	1 STANDARD GALLON
128 CUBIC FEET	1 CORD OF WOOD

TO GET CUBIC CONTENTS, MULTIPLY THE HEIGHT BY THE WIDTH BY THE LENGTH.

WEIGHT	
16 DRAMS	1 OUNCE
16 OUNCES	1 POUND
100 POUNDS	1 HUNDRED WEIGHT
2,000 POUNDS	1 SHORT TON
2,240 POUNDS	1 LONG TON

LIQUID	
4 GILLS	1 PINT
2 PINTS	1 QUART
4 QUARTS	1 GALLON
7.48 GALLONS	1 CUBIC FOOT
31-1/2 GALLONS	1 BARREL
1 GALLON	231 CUBIC INCHES

METRIC EQUIVALENTS-----

THIS TABLE WILL ENABLE YOU TO QUICKLY CONVERT METRIC MEASUREMENTS TO APPROXIMATE U.S. TERMS.

1 DECIMETER	4 INCHES
1 METER	1.1 YARDS
1 KILOMETER	5/8 MILE
1 HECTARE	2-1/2 ACRES
1 STERE	1/4 CORD
1 LITER (LIQUID)	1.06 QUART
1 LITER (DRY)	0.9 QUART
1 HECTOLITER	2-5/8 BUSHELS
1 KILOGRAM	2-1/5 POUNDS
1 METRIC TON	2,204.6 POUNDS

SEE: *METRIC MEASUREMENTS.*

METRIC MEASUREMENTS-----

A 2 X 4 WILL ALWAYS BE CALLED A TWO BY FOUR (THAT IS ITS NAME). PERHAPS IN THE FUTURE METRIC SIZES WILL CONTROL THE DIMENSIONS OF BUILDING MATERIALS AND PRODUCTS. THIS METRIC CONVERSION CHART SHOWING LENGTH, SURFACE AND VOLUME IS FOR YOUR CONVENIENCE.

LENGTH

1 METER = 39.37 INCHES	1 INCH = .0254 METERS
1 METER = 3.281 FEET	1 FOOT = .3048 METERS
1 METER = 1.093 YARDS	1 YARD = .9144 METERS

SURFACE

1 SQ. CENTIMETER = .155 SQ. INCHES	1 SQ. INCH = 6.45 SQ. CENTIMETERS
1 SQ. METER = 10.764 SQ. FEET	1 SQ. INCH = .092 SQ. METERS
1 SQ. METER = 1.196 SQ. YARDS	1 SQ. YARD = .836 SQ. METERS
1 SQ. KILOMETER = .386 SQ. MILES	1 SQ. MILE = 2.59 SQ. KILOMETERS
1 HECTARE = 2.471 ACRES	1 ACRE = .405 HECTARE

VOLUME

1 CU. CENTIMETER = .061 CU. INCH	1 CU. INCH = 16.387 CU. CENTIMTRS
1 CUBIC METER = 35.314 CU. FEET	1 CU. FOOT = .028 CUBIC METERS
1 CUBIC METER = 1.308 CU. YARDS	1 CU. YARD = .764 CUBIC METERS
1 STERE = .275 CORD	1 CORD (WOOD) = 3.624 STERES

METRIC SYSTEM-----

THE METRIC MEASUREMENT OF LENGTH, CAPACITY, WEIGHT AND AREAS IS A SYSTEM BASED VERY SIMPLY ON A MULTIPLE OF 10, SIMILAR TO OUR MONETARY SET-UP.

THE BASIC WORDS USED ARE METER FOR LENGTH AND AREA, GRAM FOR WEIGHT OR MASS, AND LITER FOR CAPACITY. THE PREFIXES TO THESE WORDS (AS SHOWN IN THE *METRIC PREFIXES* TABLE) SHOWS THE NUMBER OF BASIC UNITS. FOR EXAMPLE, 10 METERS EQUAL 1 DEKAMETER OR 10 LITERS EQUAL 1 DEKALITER. SEE THE *METRIC TABLES AND ABBREVIATIONS* WHICH COVER THE VARIOUS FORMS OF MEASUREMENT IN THIS SYSTEM.

METRIC TABLES AND ABBREVIATIONS-----

LINEAR	
10 MILLIMETERS (MM)	= 1 CENTIMETER (CM)
10 CENTIMETERS	= 1 DECIMETER (DM)
10 DECIMETERS	= 1 METER (M)
10 METERS	= 1 HECTOMETER (HM)
10 HECTOMETERS	= 1 KILOMETER (KM)

AREAS (SQUARE)	
100 SQUARE MILLIMETERS (MM^2)	= 1 SQUARE CENTIMETER (CM^2)
10,000 SQUARE CENTIMETERS	= 1 SQUARE METER (M^2)
100 SQUARE METERS	= 1 ARE (A)
100 ARES	= 1 HECTARE (HA)
100 HECTARES	= 1 SQUARE KILOMETER (KM^2)

VOLUME OR CAPACITY (LIQUID)	
10 MILLILITERS (ML)	= 1 CENTILITER (CL)
10 CENTILITERS	= 1 DECILITER (DL)
10 DECILITERS	= 1 LITER (L)
10 LITERS	= 1 DEKALITER (DAL)
10 DEKALITERS	= 1 HECTOLITER (HL)
10 HECTOLITERS	= 1 KILOLITER (KL)

CUBIC METERS	
1,000 CUBIC MILLIMETERS (MM^3)	= 1 CUBIC CENTIMETER (CM^3)
1,000 CUBIC CENTIMETERS	= 1 CUBIC DECIMETER (DM^3)
1,000 CUBIC DECIMETERS	= 1 CUBIC METER (M^3)
1 CUBIC METER	= 1 STERE

WEIGHT	
10 MILLIGRAMS (MG)	= 1 CENTIGRAM (CG)
10 CENTIGRAMS	= 1 DECIGRAM (DG)
10 DECIGRAMS	= 1 GRAM (G)
10 GRAMS	= 1 DEKAGRAM (DAG)
10 DEKAGRAMS	= 1 HECTOGRAM (HG)
10 HECTOGRAMS	= 1 KILOGRAM (KG)
1,000 KILOGRAMS	= 1 METRIC TON (T)

LIQUID CONVERSION	
1 PINT LIQ.	0.473 LITER
1 PINT DRY	0.551 LITER
1 QUART LIQ.	0.946 LITER
1 QUART DRY	1.101 LITERS
1 GALLON LIQ.	3,785 LITERS

SEE: *METRIC MEASUREMENTS, MEASUREMENTS u S.*

METRIC PREFIXES-----

PREFIX	MEANING
MILLI	1,000 TH
CENTI	100 TH
DECI	10 TH
DEKA	10
HECTO	100
KILO	1,000

MULTIPLICATION TABLE-----

IF YOU DO NOT HAVE READY ACCESS TO A CALCULATOR YOU CAN SAVE TIME BY USING THIS SIMPLE TABLE TO FIND THE AREA OF FLOORS, WALLS, CEILINGS. YOU CAN ALSO USE IT FOR ANY OTHER PROBLEMS WHERE YOU WANT TO MULTIPLY FIGURES.

TO USE THE TABLE SIMPLY LOOK IN THE LEFT COLUMN B FOR ONE DIMENSION. THEN FIND THE OTHER DIMENSION ON THE TOP LINE A. WHERE THE TWO LINES INTERSECT, YOU WILL FIND THE ANSWER.

FOR EXAMPLE WHERE LINE 7 ON THE LEFT COLUMN INTERSECTS LINE 7 ON THE TOP LINE YOU WILL FIND THE FIGURE 49.

YOU CAN USE THE TABLE FOR QUICK CALCULATIONS OF GREATER DIMENSIONS BY ADDING THE RESULTS FOR TWO OR MORE CALCULATIONS.

	A														
B	1	2	3	4	5	6	7	8	9	10	11	12	13	14	15
	2	4	6	8	10	12	14	16	18	20	22	24	26	28	30
	3	6	9	12	15	18	21	24	27	30	33	36	39	42	45
	4	8	12	16	20	24	28	32	36	40	44	48	52	56	60
	5	10	15	20	25	30	35	40	45	50	55	60	65	70	75
	6	12	18	24	30	36	42	48	54	60	66	72	78	84	90
	7	14	21	28	35	42	49	56	63	70	77	84	91	98	105
	8	16	24	32	40	48	56	64	72	80	88	96	104	112	120
	9	18	27	36	45	54	63	72	81	90	99	108	117	126	135
	10	20	30	40	50	60	70	80	90	100	110	120	130	140	150
	11	22	33	44	55	66	77	88	99	110	121	132	143	154	165
	12	24	36	48	60	72	84	96	108	120	132	144	156	168	180
	13	26	39	52	65	78	91	104	117	130	143	156	169	182	195
	14	28	42	56	70	84	98	112	126	140	154	168	182	196	210
	15	30	45	60	75	90	105	120	135	150	165	180	195	210	225

NAILS-----

THIS TABLE WILL ENABLE YOU TO SELECT THE PROPER SIZE COMMON OR FINISH NAILS FOR VARIOUS USES IN BUILDING, REPAIRING, REMODELING OR RESTORATION WORK.

COMMON NAILS

Size*	Length (inches)	Diameter gauge number	Diameter of head (inches)	Approx-imate number per pound
2d	1	15	11/64	830
3d	1-1/4	14	13/64	528
4d	1-1/2	12-1/2	1/4	316
5d	1-3/4	12-1/2	1/4	271
6d	2	11-1/2	17/64	168
7d	2-1/4	11-1/2	17/64	150
8d	2-1/2	10-1/4	9/32	106
9d	2-3/4	10-1/4	9/32	96
10d	3	9	5/16	69
12d	3-1/4	9	5/16	63
16d	3-1/2	8	11/32	49
20d	4	6	13/32	31
30d	4-1/2	5	7/16	24
40d	5	4	15/32	18
50d	5-1/2	3	1/2	14
60d	6	2	17/32	11

FINISHING NAILS

Size*	Length (inches)	Diameter gauge number	Diameter of head gauge number	Approx-imate number per pound
2d	1	16-1/2	13-1/2	1,351
3d	1-1/4	15-1/2	12-1/2	807
4d	1-1/2	15	12	584
5d	1-3/4	15	12	500
6d	2	13	10	309
8d	2-1/2	12-1/2	9-1/2	189
10d	3	11-1/2	8-1/2	121
16d	3-1/2	11	8	90
20d	4	10	7	62

*The nail industry still clings to the penny system to indicate the length of the most commonly used nails, ranging in length from 1 to 6 inches. The symbol for penny is d.

COURTESY BUILD IT WITH CLIPS
(THE PANEL CLIP CO.)

NAIL TYPES----

COMMON AND FINISH NAILS ARE USED FOR MOST TYPES OF BUILDING WORK. IN THE SKETCH BELOW ARE SEVERAL OTHER TYPES OF NAILS THAT MAY BE REQUIRED FOR INSTALLING SPECIAL MATERIALS OR PRODUCTS.

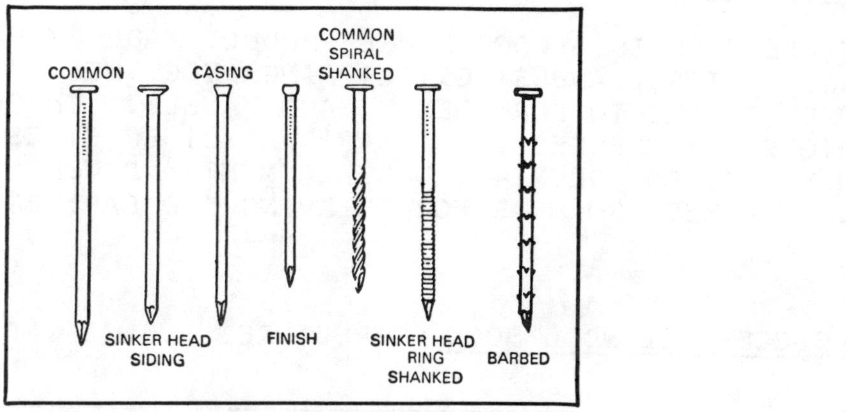

COURTESY BUILD IT WITH CLIPS
(THE PANEL CLIP CO.)

SEE: SUCCESSFUL PUTTING IT ALL TOGETHER AND SUCCESSFUL HOME-OWNER'S TOOLS (STRUCTURES PUBLISHING CO.)

NAILING-----

AS SHOWN IN THE SKETCHES BELOW, NAILING VARIOUS KINDS OF WOOD REQUIRES DIFFERENT NAIL LENGTHS. THE VARIOUS TYPES OF JOINTS AND NAILING METHODS ARE ALSO SHOWN HERE.

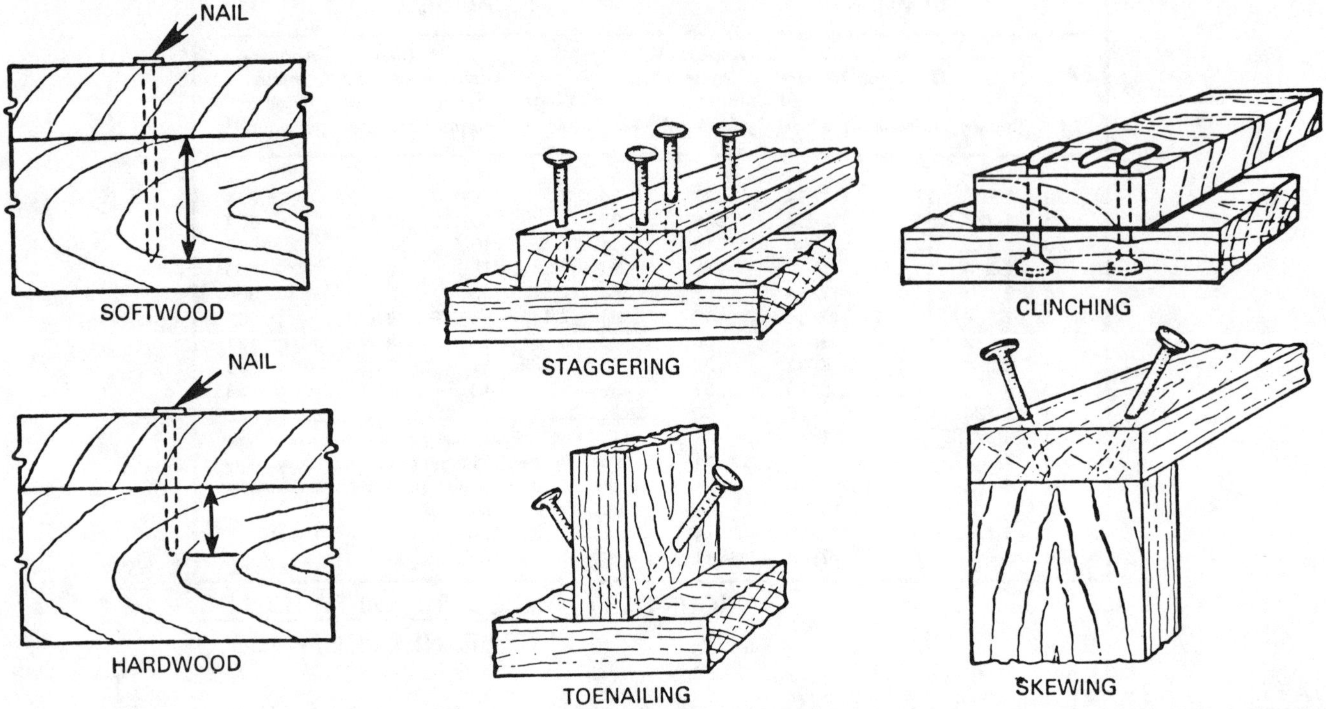

COURTESY BUILD IT WITH CLIPS (THE PANEL CLIP CO.)

SEE: HOW TO BUILD YOUR OWN HOME (STRUCTURES PUBLISHING COMPANY)

PARTICLEBOARD-----

PARTICLEBOARD IS A WOOD PANEL PRODUCT MADE BY PRESSING WOOD SHAVINGS, CHIPS, SAWDUST OR CLIPPINGS TOGETHER. IT IS THEN BONDED BY RESINS TO FORM DENSE HARD PANELS. THESE PANELS COME IN VARIOUS THICKNESSES IN 4' X 8' AND OTHER SIZES. THEY ARE COMMONLY USED FOR DECKING, UNDERLAYMENT AND NUMEROUS OTHER PLACES IN HOMES, SUCH AS FOR SHELVING AND CABINETS.

SEE: SUCCESSFUL WOOD BOOK (STRUCTURES PUBLISHING COMPANY)

Notes: ______________________________

____

____

____

____

____

PLOT SIZES-----

WHILE MOST LOTS AND TRACTS USUALLY SHOW THE EXACT SIZE AND AREA, OCCASIONALLY YOU MAY CONSIDER BUYING OR SELLING ODD SIZED SITES BEFORE SURVEYS ARE MADE. IN THIS CASE YOU CAN USE THIS SIMPLE METHOD TO APPROXIMATE THE SIZE OF THE PLOT.

TO CALCULATE THE APPROXIMATE AREA OR ACREAGE OF A PLOT OR PLAT THAT IS NOT SQUARE OR A UNIFORM RECTANGLE, USE THE FOLLOWING METHOD. DRAW THE AREA ON PAPER TO SCALE OR MEASURE OFF THE SITE INTO SQUARES, UNIFORM RECTANGLES AND TRIANGLES WITH RIGHT ANGLES. COMPUTE THE NUMBER OF SQUARE FEET IN EACH SECTION. THIS IS DONE BY MULTIPLYING THE LENGTH BY THE WIDTH TO GET THE NUMBER OF SQUARE FEET IN BOTH SQUARES AND RECTANGLES. TO GET THE TOTAL SQUARE FOOTAGE IN THE TRIANGLES, (RIGHT ANGLED), MULTIPLY ONE HALF THE LENGTH BY THE WIDTH. WHEN YOU GET THE TOTAL NUMBER OF SQUARE FEET IN THE AREA, DIVIDE BY 43,560 (NUMBER OF SQUARE FEET IN AN ACRE) TO GET THE APPROXIMATE ACREAGE. FOR EXAMPLE, A LOT 60 FEET BY 120 FEET WOULD CONTAIN 7,200 SQUARE FEET OR WHEN DIVIDED BY 43,560 WOULD EQUAL .1652 ACRES.

BEFORE DEEDS CAN BE MADE AND EXECUTED, SURVEYS OR EXACT DESCRIPTIONS OF ALL PROPERTY MUST BE MADE AND RECORDED FOR TAX PURPOSES.

SEE: *ACREAGE CALCULATIONS*.

PROPERTY LINES-----

THIS INFORMATION IS SHOWN ON MAPS OF THE AREA WHICH CAN BE OBTAINED FROM THE ENGINEERING OR MAP DEPARTMENT OF THE CITY, COUNTY OR DIVISION HANDLING THESE MATTERS. EXPERIENCED PERSONNEL ARE USUALLY AVAILABLE TO ASSIST IN LOCATING PROPERTY MAPS. A COPY OF THE MAP SHEET OR SHEETS SHOWING THE PROPERTY IN QUESTION USUALLY IS AVAILABLE ON SHORT NOTICE AT A NOMINAL FEE.

IN CASES WHERE A SURVEY IS REQUIRED BY THE BUYER OR A FINANCING INSTITUTION, THE COST INCURRED SHOULD BE COVERED IN THE SALES AGREEMENT. THE MAPS AND PLOT PLANS ON FILE ARE USED TO DETERMINE THE BOUNDARIES AS DESCRIBED IN THE DEED FOR THE PROPERTY. THESE MAPS SHOW THE LOCATION OF STREETS, ROADS, SEWERS, ETC., AND USUALLY INDICATE THE EASEMENT FOR UNUSUAL CONDITIONS AFFECTING THE PROPERTY.

Notes:

PLYWOOD-----

PLYWOOD IS MADE FROM A SERIES OF THIN SHEETS OF VENEER PEELED FROM LOGS AND BONDED TOGETHER BY HEAT AND PRESSURE TO FORM PANELS. AS SHOWN IN THE SKETCH AT THE RIGHT, THE PLIES RUN AT RIGHT ANGLES TO EACH OTHER TO GIVE STRENGTH AND RESIST STRESSES IN ALL DIRECTIONS.

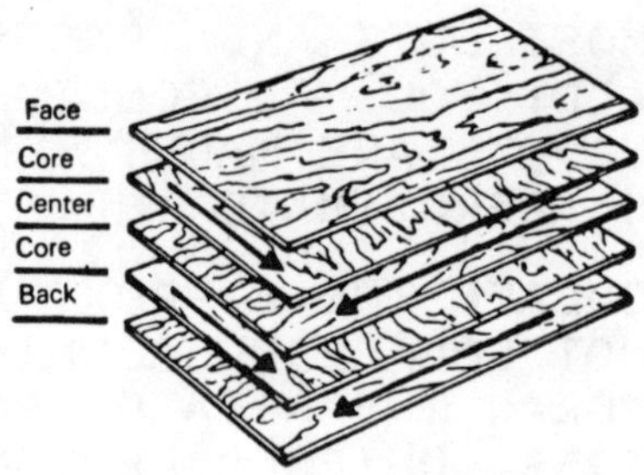

PLYWOOD OF VARIOUS TYPES AND GRADES IS USUALLY INDICATED BY A STAMP ON THE PANELS. THIS MARKING INDICATES THE GRADE AND TYPE AND THE INTENDED USE, WHETHER INTERIOR OR EXTERIOR. THE GRADES GENERALLY AVAILABLE ARE "A" THROUGH "D."

A. SMOOTH SURFACE THAT CAN BE PAINTED. USED FOR NATURAL FINISH IN SOME CASES.

B. A SOLID SURFACE VENEER. SOME SMALL KNOTS AND CIRCULAR REPAIR PLUGS ARE ALLOWED.

C. LIMITED SPLITS AND UP TO 1" KNOT HOLES ALLOWED. ALSO PLUGGED, WITH UP TO 1/8" SPLITS AND PLUGGED KNOT HOLES.

D. KNOTS TO 2-1/2" AND LIMITED SPLITS.

PLYWOOD WALL PANELING-----

INTERIOR WALL PANELING OF PLYWOOD IS NOW USED VERY WIDELY IN NEW WORK AND REMODELING. IT IS ESPECIALLY ADAPTABLE FOR USE IN FAMILY AND RECREATION ROOMS, DENS AND OTHER AREAS OF THE HOME. IT IS MADE IN 1/8", 5/32", 3/16", 1/4" AND 7/16" THICKNESSES AND IN 4'X8' SHEETS. MANY OF THESE PANELS HAVE VENEER FACINGS OF SOFT WOODS AND HARDWOOD WITH MANY SURFACE TREATMENTS, COLORS AND PATTERNS.

AS SHOWN IN THE LEFT SKETCH BELOW, SOME TYPES ARE GROOVED IN RANDOM WIDTHS OR PLANKED EFFECTS ON 16" CENTERS FOR NAILING TO STUDS. THE RIGHT SKETCH SHOWS CROSS SCORING TO GIVE A PLANED EFFECT AND SOME OF THE DIFFERENT TYPE GROOVES. GROOVES ARE GENERALLY STRIPED DARKER THAN THE PANEL SURFACE.

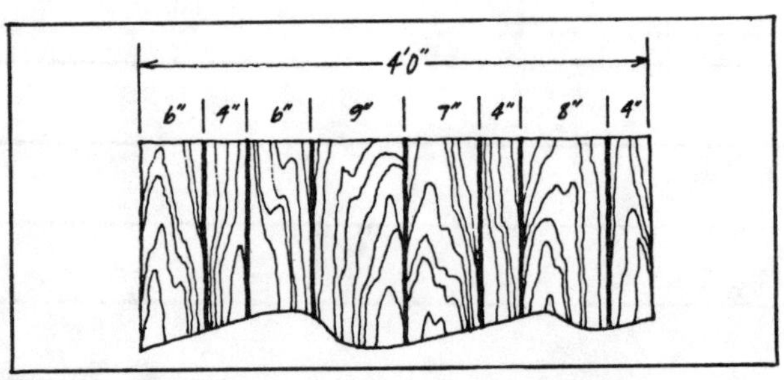

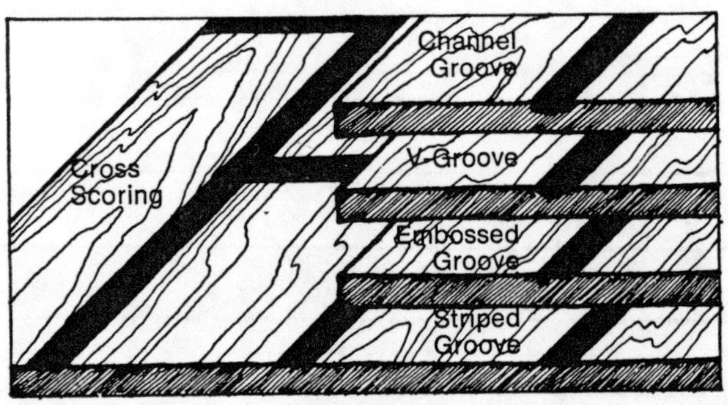

PLYWOOD SIDING-----

SOME TYPES OF PLYWOOD ARE PREPARED FOR EXTERIOR USE AND MAY BE FOUND IN SEVERAL TYPES AND VARIETIES OF WOOD. COMMON TYPES FOR SIDING ARE PLAIN PANELS, ROUGH SAWED, ROUGH SAWED TEXTURED, AND ONE ELEVEN ROUGH SAWED OR TEXTURED—WHICH HAS NARROW VERTICAL GROOVES ON 4" AND 8" CENTERS. ANOTHER TYPE, THE REVERSE BOARD AND BATTEN DESIGN, HAS GROOVES CUT ON 8" AND 12" CENTERS, AND IS ALSO MADE IN ROUGH SAWED FINISH. MOST TYPES ARE MADE IN 8' AND 9' LENGTHS 4' WIDE.

POINTS, LOAN-----

POINTS ARE A COMMONLY USED TERM TO DESCRIBE THE PAYMENT OF SPECIAL FEES WHEN OBTAINING OR MAKING LOANS. IT IS USUALLY BASED ON A PERCENTAGE OF THE TOTAL LOAN AND MAY BE IN ADDITION TO A LEGITIMATE SERVICE FEE USUALLY CHARGED BY LENDING INSTITUTIONS. TWO POINTS, FOR EXAMPLE, WOULD BE 2 PERCENT OF THE LOAN. THUS 2 POINTS ON A $20,000 LOAN WOULD BE $400.

PRODUCT BUREAUS-----

THE MANUFACTURERS AND PRODUCERS OF MANY BUILDING PRODUCTS AND MATERIALS SPONSOR AND MAINTAIN BUREAUS FOR RESEARCHING AND TESTING THEIR PRODUCTS AND ADAPTING THEM TO THE BEST POSSIBLE USE.

YOU CAN BENEFIT BY UTILIZING THE MANY HELPFUL IDEAS AND SUGGESTIONS PRESENTED BY THESE BUREAUS. THE LITERATURE THEY DEVELOP AND PRODUCE IS FREE. IN ADDITION, THE FIRMS WHO PRODUCE MANY COMMONLY USED MATERIALS PREPARE AND FURNISH HELPFUL LITERATURE.

REAL ESTATE ADS-----

IF YOU ARE BUYING OR SELLING REAL PROPERTY YOU MAY WANT TO USE REAL ESTATE CLASSIFIED NEWSPAPER ADVERTISING. THESE ADS USUALLY ARE SOLD BY THE COLUMN INCH OR BY THE WORD.

IN THE INTEREST OF ECONOMY, YOU WILL FIND OR USE MANY REAL ESTATE ABBREVIATIONS COVERING ALMOST ALL THE USUAL TERMS USED IN THIS BUSINESS.

TO HELP YOU UNDERSTAND REAL ESTATE ADS OR TO ASSIST YOU IN PREPARING THEM, REFER TO THE COMPLETE LIST OF *REAL ESTATE TERMS AND ABBREVIATIONS* FOUND IN THIS BOOK.

SEE: *BUYING AND SELLING REAL ESTATE, REAL ESTATE TERMS.*

PLUMBING TERMS-----

HERE IS A LIST OF COMMON TERMS USED BY THE PLUMBING INDUSTRY IN CONNECTION WITH HOUSE CONSTRUCTION.

- ACCEPTED STANDARDS - STANDARDS APPROVED BY THE AUTHORITY HAVING JURISDICTION OVER PLUMBING.
- APPROVED - ACCEPTED AS SATISFACTORY BY THE AUTHORITY HAVING JURISDICTION OVER PLUMBING.
- BUILDING-DRAINAGE SYSTEM - THIS CONSISTS OF ALL PIPING PROVIDED FOR CARRYING WASTE WATER, SEWAGE, OR OTHER DRAINING FROM THE BUILDING TO THE STREET SEWER OR PLACE OF DISPOSAL.
- BUILDING SEWER - PART OF THE HORIZONTAL PIPING OF A BUILDING-DRAINAGE SYSTEM THAT EXTENDS FROM THE BUILDING DRAIN 5 FEET OUTSIDE OF THE INNER FACE OF THE BUILDING WALL TO THE STREET SEWER OR OTHER PLACE OF DISPOSAL; CONVEYS THE DRAINAGE OF ONE BUILDING SITE.
- DRAIN - A DRAIN OR DRAIN PIPE IS ANY PIPE WHICH CARRIES WATER OR WATER-BORNE WASTES IN A BUILDING-DRAINAGE SYSTEM.
- DRY VENT - ANY VENT THAT DOES NOT CARRY WATER OR WATER-BORNE WASTES.
- FIXTURE BRANCH - THE SUPPLY PIPE BETWEEN THE FIXTURE AND THE WATER DISTRIBUTING PIPE.
- FIXTURE DRAIN - THE DRAIN FROM THE TRAP OF A FIXTURE TO THE JUNCTION OF THE DRAIN WITH ANY OTHER DRAIN PIPE.
- FIXTURE UNIT - A FACTOR CHOSEN SO THAT THE LOAD PRODUCING VALVES OF THE DIFFERENT PLUMBING FIXTURES CAN BE EXPRESSED APPROXIMATELY AS MULTIPLES OF THAT FACTOR.
- GRADE - THE GRADE OF A LINE OF PIPE IS ITS SLOPE. IN PLUMBING IT IS THE FALL IN INCHES PER FOOT.
- MAIN - THE PRINCIPAL ARTERY OF THE PLUMBING SYSTEM TO WHICH BRANCHES MAY BE CONNECTED.
- PLUMBING FIXTURES - RECEPTACLES WHICH RECEIVE AND DISCHARGE WATER, OR WATER-BORNE WASTES INTO A DRAINAGE SYSTEM.
- PLUMBING SYSTEM - INCLUDES THE WATER SUPPLY DISTRIBUTING PIPES, THE FIXTURES AND FIXTURE TRAPS, THE SOIL, WASTE, AND VENT PIPES, THE BUILDING DRAIN AND BUILDING SEWER, AND THE STORM DRAINAGE PIPES.
- RISER - A WATER SUPPLY PIPE WHICH EXTENDS VERTICALLY ONE FULL STORY OR MORE TO CONVEY WATER TO BRANCHES OR FIXTURES.

PLUMBING TERMS (continued)

- SANITARY SEWER - A SEWER DESIGNED OR USED ONLY FOR CONVEYING LIQUID OR WATER-BORNE WASTE FROM PLUMBING FIXTURES.
- SOIL PIPE - ANY PIPE WHICH CARRIES THE DISCHARGE OF WATER CLOSETS OR OTHER FIXTURES.
- STACK - THE TERM FOR THE VERTICAL MAIN OF SOIL, WASTE OR VENT PIPING.
- TRAP - A FITTING OR DEVICE DESIGNED TO PROVIDE A LIQUID TRAP SEAL THAT PREVENTS THE PASSAGE OF AIR.
- VENT - A PIPE TO PROVIDE A FLOW OF AIR TO OR FROM A DRAINAGE SYSTEM AND TO PROTECT TRAP SEALS FROM SIPHONAGE AND BACK PRESSURE.
- WASTE PIPE - A DRAIN PIPE WHICH RECEIVES THE DISCHARGE OF ANY FIXTURE RECEIVING HUMAN WASTE.
- WATER MAIN - A WATER SUPPLY PIPE FOR PUBLIC OR COMMUNITY USE.
- WATER SUPPLY SYSTEM - CONSISTS OF THE WATER SERVICE PIPE, THE WATER DISTRIBUTING PIPES, AND THE NECESSARY CONNECTING PIPES, FITTINGS, AND CONTROL VALVES.

PIPE MEASUREMENTS-----

WHEN MEASURING FOR THREADED PIPE INSTALLATION OR REPLACEMENT THE SKETCH BELOW WILL BE HELPFUL WHEN BUYING PIPE FITTINGS.

ALWAYS USE CENTER TO CENTER MEASUREMENTS FOR PIPE. IF FITTINGS ARE REQUIRED USE CENTER TO FACE DIMENSIONS.

THE SKETCH SHOWS:

X: CENTER-TO-CENTER MEASUREMENTS. A: CENTER-TO-FACE DIMENSIONS.

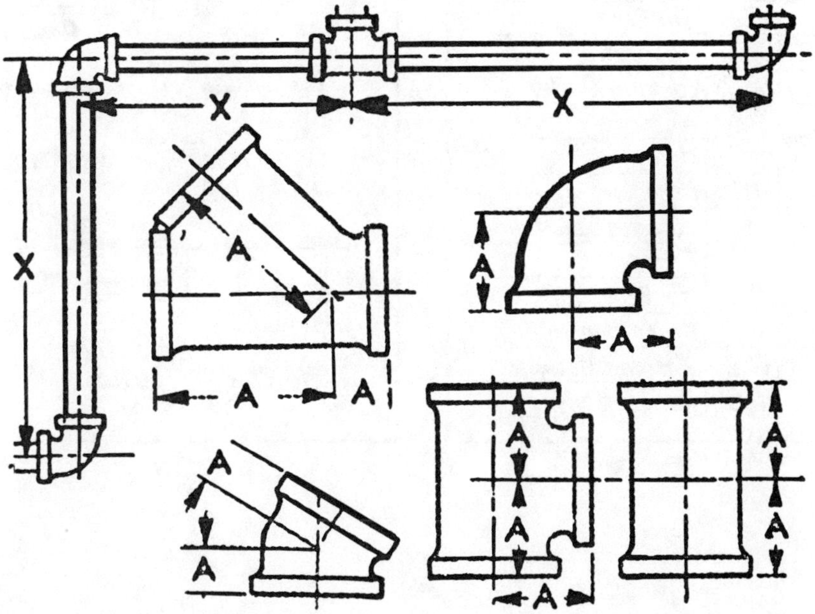

PLUMBING SYMBOLS-----

THIS GROUP OF PLUMBING SYMBOLS WILL HELP YOU BETTER UNDERSTAND THE PLUMBING LAYOUT ON PLANS FOR HOMES AND RENOVATION.

THE SKETCHES BELOW WILL ENABLE YOU TO IDENTIFY AND RECOGNIZE THE VARIOUS TYPES AND KINDS OF PLUMBING FIXTURES AND EQUIPMENT.

MANY OTHER TYPES AND SIZES (SOME IMPORTED) MAY BE FOUND IN A SELECTION OF COLORS TO MATCH MOST DECOR ARRANGEMENTS.

IN ADDITION TO THE LAVATORY FIXTURES SHOWN BELOW, DROP-IN COUNTER TYPES ARE USED WITH AND FITTED TO VANITY LAVATORY CABINETS OF VARIOUS WIDTHS.

Symbol	Plan	Initials	Item
		D.	Drainage Line
		V.S.	Vent Line
			Tile Pipe
		C.W.	Cold Water Line
		H.W.	Hot Water Line
		H.W.R.	Hot Water Return
		G	Gas Pipe
		D.W.	Ice Water Supply
		D.R.	Ice Water Return
		F.L.	Fire Line
		I.W.	Indirect Waste
		I.S.	Industrial Sewer
		AW	Acid Waste
	A	A	Air Line
	V	V	Vacuum Line
	R	R	Refrigerator Waste
			Gate Valves
			Check Valves
CO CO		CO.	Cleanout
F.D		F.D.	Floor Drain
R.D		R.D.	Roof Drain
REF.		REF.	Refrigerator Drain
		S.D.	Shower Drain
G.T.		G.T.	Grease Trap
S.C.		S.C.	Sill Cock
G		G.	Gas Outlet
VAC		VAC.	Vacuum Outlet
M		M	Meter
I×			Hydrant
H.R.		H.R	Hose Rack
H.R.		H.R.	Hose Rack-Built in
L		L.	Leader
H WT		H.W.T.	Hot Water Tank
WH		W.H.	Water Heater
WM		W.M	Washing Machine
RB		R.B.	Range Boiler

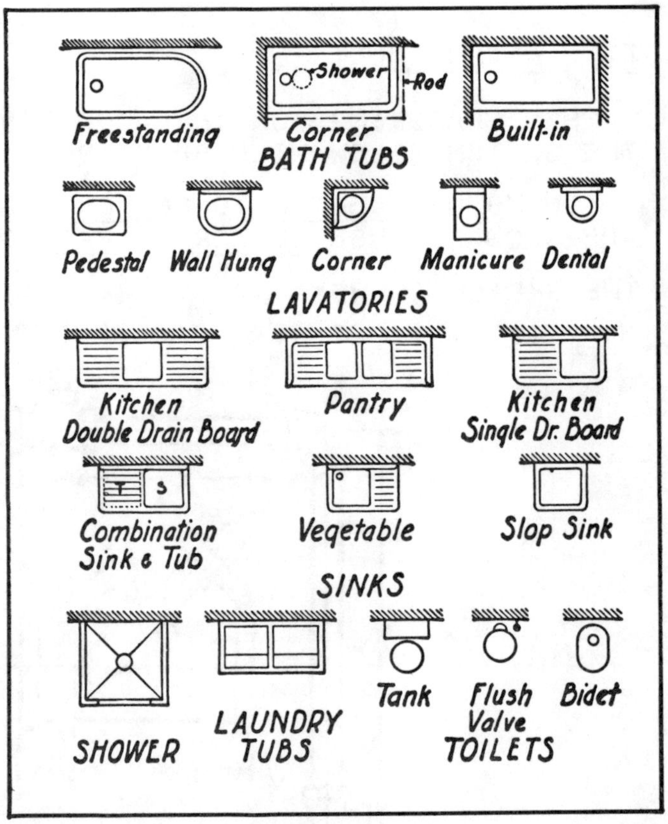

PLOT PLAN-----

USUALLY WHEN AN APPLICATION IS MADE FOR A BUILDING PERMIT FOR A NEW HOME OR EXTENSIVE ALTERATIONS, A PLOT PLAN IS REQUIRED. THIS IS THE PLAN OF THE PROPERTY AND THE PROPOSED LOCATION OF THE NEW WORK. THEY ARE MADE ON AN ENGINEER'S SCALE OF 50, 100 OR 200 FEET PER INCH, AS COMPARED TO HOUSE PLANS WHICH ARE MADE ON AN ARCHITECT'S SCALE OF 1/4" PER INCH.

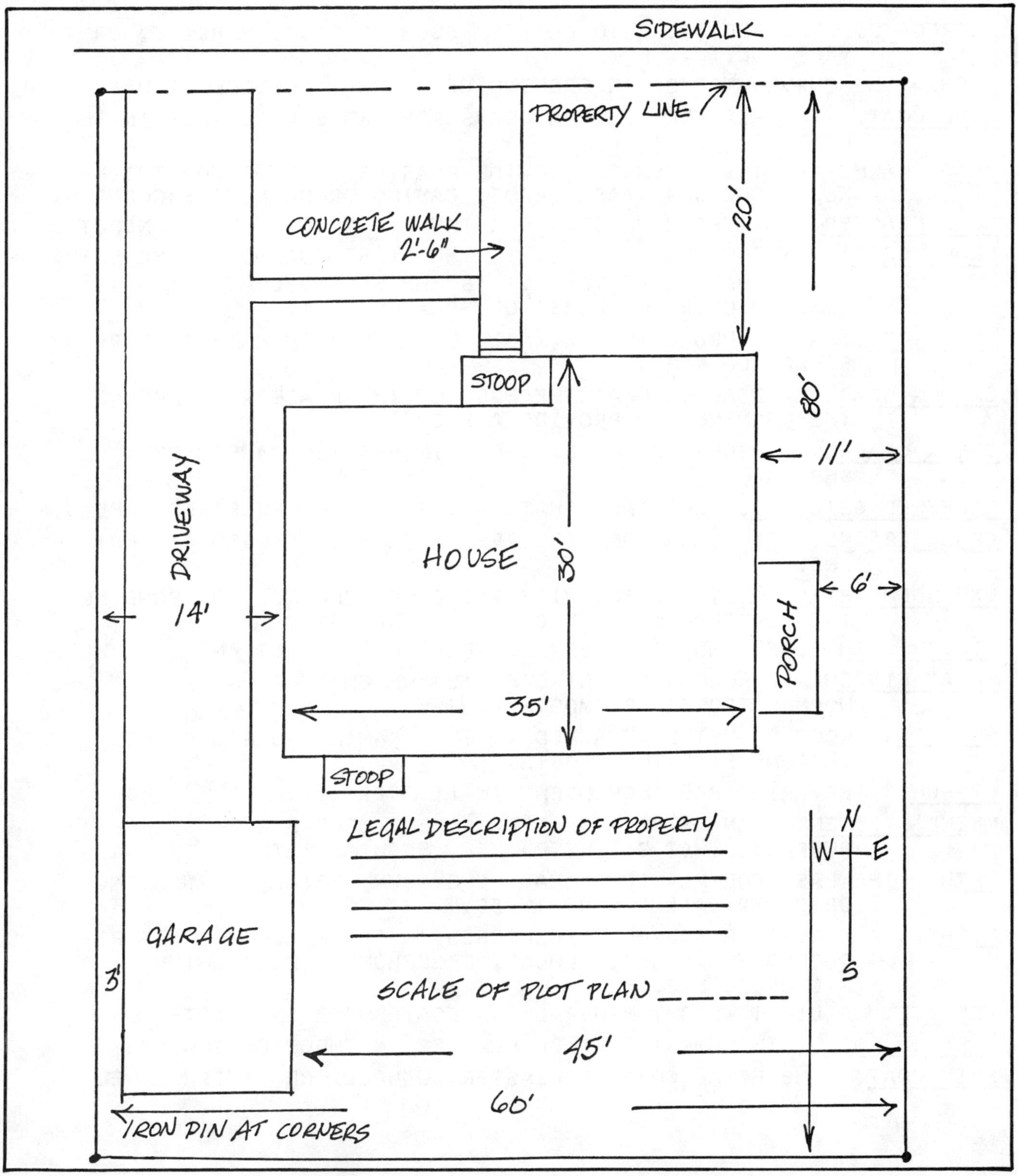

PLASTER TERMS-----

PLASTER IS STILL USED ON SOME NEW JOBS AND IS OFTEN USED IN RENOVATION OR REMODELING PROJECTS TO MATCH WALLS. THESE ARE SOME OF THE COMMON TERMS USED IN PLASTERING.

ACOUSTICAL PLASTER. FINISHING PLASTERED SURFACES TO CUT DOWN NOISE.

AGGREGATE. THE ADDITIONS TO PLASTER SUCH AS SAND, PERLITE, OR VERMICULITE.

ARCH. THE CURVED TOP OF AN OPENING OF DOORWAY, WINDOW OR HALL.

BASE COAT. THE PLASTER COAT OR COATS APPLIED BEFORE THE FINISH COAT.

BEAD. A METAL STRIP USED TO PROVIDE PLASTER GROUNDS FOR THICKNESS, CORNERS, BASE BEADS, CASING OR BULL NOSE CORNERS.

BOND PLASTER. A GYPSUM PLASTER FOR THE FIRST COAT OVER CONCRETE.

BROWN COAT. IN TWO COAT WORK IT IS THE FIRST COAT OVER THE LATH. IN THREE COAT WORK IT IS THE SECOND COAT WHICH IS APPLIED OVER THE FIRST OR SCRATCH COAT.

BULL NOSE. A ROUNDED EXTERNAL ANGLE USED TO ELIMINATE A SHARP OR SQUARE CORNER.

CASING BEAD. A BEAD USED AT THE EDGE OF THE PLASTER OR AROUND THE OPENINGS TO PROVIDE A STOP.

CHECK CRACKS. CHECKS OR SMALL CRACKS IN PLASTER CAUSED BY SHRINKAGE.

COFFERED CEILINGS. CEILINGS THAT HAVE SUNKEN OR RECESSED PANELS.

CRAZE CRACKS. FINE, RANDOM FISSURES IN PLASTER CAUSED BY SHRINKING.

EXPANDED METAL. METAL SHEET WITH SLITS OR OPENINGS FOR BONDING THE PLASTER—USUALLY CALLED METAL LATH.

FINISH. THE LAST AND FINAL COAT OF PLASTERED WALLS AND CEILINGS.

FLOAT FINISH. A ROUGH APPEARING FINISH WHICH IS ROUGHER THAN THE NORMAL TROWEL OR SMOOTH FINISH.

GROUNDS. WOOD OR METAL ATTACHED TO THE FRAMING TO GAUGE THE THICKNESS OF THE PLASTER.

GYPSUM. THE PRINCIPAL INGREDIENT IN PLASTERS.

KEENE'S CEMENT. AN EXTRA HARD TYPE CEMENT USED FOR PLASTER SURFACES THAT GET EXTRA USE, SUCH AS BATHS.

LATH. THE BASE FOR PLASTER. MAY BE GYPSUM, METAL, INSULATING OR PAPER-BACKED WIRE TYPES.

NICHE. A CURVED OR SQUARE TYPE RECESS IN THE WALL—USED AS REQUIRED FOR VASES, CLOCKS, TELEPHONES, DOOR CHIMES, PICTURES, ETC.

PERLITE. A LIGHT WEIGHT MINERAL AGGREGATE USED IN PLASTER.

SCRATCH COAT. THE FIRST COAT OF PLASTER IN THREE COAT WORK.

SKIM COAT. THE FINAL COAT OF PLASTER, USUALLY OF FINISH LIME.

PLASTER TERMS *(continued)*

STUCCO. THIS DENOTES PLASTER WHEN USED ON EXTERIOR SURFACES.

VERMICULITE. EXPANDED MICA USED AS PLASTER AGGREGATE DUE TO ITS LIGHT WEIGHT AND INSULATING QUALITIES.

WAINSCOT. THE LOWER PART OF A FINISHED INTERIOR WALL—FINISHED DIFFERENTLY FROM THE UPPER PART OF THE WALL.

WHITE COAT. A LIME GAUGED PLASTER MIX USED FOR TROWEL FINISHES.

WOOD FIBRE. WOOD PARTICLES OF NON-STAINING QUALITY USED AS AN AGGREGATE WITH GYPSUM PLASTER.

PAINT COLORS, REFLECTION VALUES-----

APPLE GREEN	30%
BLACK	2%
BROWN (LIGHT)	20%
BUFF	83%
BUFF (LIGHT)	74%
CAEN STONE	76%
CANARY YELLOW	77%
COCONUT BROWN	16%
CORAL	64%
CREAM	77%
CREAM GRAY	66%
CREAM (LIGHT)	77%
FOREST GREEN	72%
FRENCH BLUE	47%
GRAY (LIGHT)	52%
GREEN (PALE)	59%
GREEN (PASTEL)	59%
IVORY	82%
IVORY TAN	66%
OLIVE TAN	43%
ORCHID	67%
ORCHID (LIGHT)	79%
OYSTER WHITE	77%
SAGE (BRIGHT)	52%
SEA GREEN	67%
SHELL PINK	55%
SILVER GRAY	46%
SKY BLUE	65%
SUNLIGHT YELLOW	77%
WHITE (SEMI-GLOSS)	80%

PAINT, COLOR MIXING-----

BUFF	WHITE, YELLOW OCHRE AND RED
CHESTNUT	RED, BLACK AND YELLOW
CHOCOLATE	RAW UMBER, RED AND BLACK
CLARET	RED, UMBER AND BLACK
COPPER	RED, YELLOW AND BLACK
DOVE	WHITE, VERMILLION, BLUE AND YELLOW
DRAB	WHITE, YELLOW OCHRE, RED AND BLACK
FAWN	WHITE, YELLOW AND RED
FLESH	WHITE, YELLOW OCHRE AND VERMILLION
FREESTONE	RED, BLACK, YELLOW OCHRE AND WHITE
FRENCH GRAY	WHITE, PRUSSIAN BLUE AND LAKE
GRAY	WHITE LEAD AND BLACK
GOLD	WHITE, STONE OCHRE AND RED
GREENE BRONZE	CHROME GREEN, BLACK AND YELLOW
GREEN PEA	WHITE AND CHROME GREEN
LEMON	WHITE AND CHROME YELLOW
LIMESTONE	WHITE, YELLOW, OCHRE, BLACK AND RED
OLIVE	YELLOW, BLUE, BLACK AND WHITE
ORANGE	YELLOW AND RED
PEACH	WHITE AND VERMILLION
PEARL	WHITE, BLACK AND BLUE
PINK	WHITE, VERMILLION AND LAKE
PURPLE	VIOLET WITH MORE RED AND WHITE
ROSE	WHITE AND MADDER LAKE
VIOLET	RED, BLUE AND WHITE

SEE: *PAINT, REFLECTION VALUES.*

STUCCO-----

YOU CAN USE STUCCO OVER SEVERAL TYPES OF MATERIALS AND CONSTRUCTION. THIS SKETCH OF THREE COAT STUCCO OVER CONCRETE MASONRY WILL SERVE TO ACQUAINT YOU WITH THE VARIOUS TERMS. STUCCO IS A CEMENT PLASTER MIXED AT THE JOB SITE: THE FINISH COAT CAN BE TINTED WITH VARIOUS CEMENT COLORING MATERIALS AND/OR THE FINISH COAT CAN BE TROWEL OR FLOAT FINISHED IN A VARIETY OF PATTERNS OR DESIGNS.

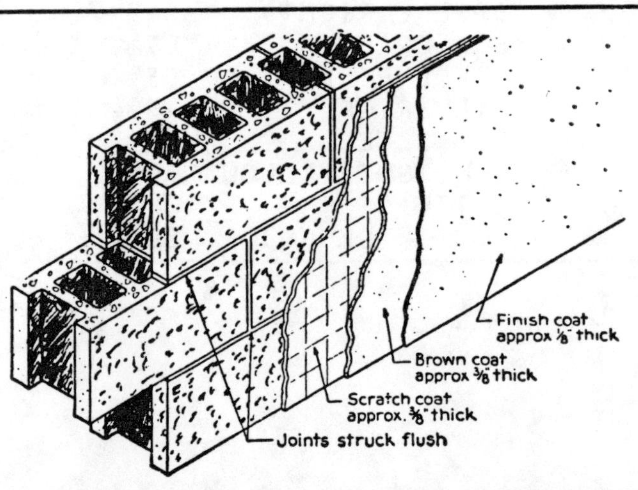

PAINT, EXTERIOR TYPES, USES-----

USING THE PROPER TYPE AND KIND OF PAINT ON EXTERIOR AREAS AND SURFACES HELPS KEEP OVERALL MAINTENANCE COSTS DOWN, PROVIDES A BETTER APPEARANCE AND PROLONGS THE LIFE OF THE TREATED AREAS.

THIS CHART SUGGESTS THE PAINT PRODUCTS AND MATERIALS BEST SUITED FOR VARIOUS EXTERIOR CONDITIONS.

FINISH TYPES / Surface Types	Oil or Oil-Alkyd Paint	Cement Powder Paint	Exterior Clear Finish	Aluminum Paint	Wood Stain	Roof Coating	Trim Paint	Porch & Deck Paint	Primer or Undercoater	Metal Prime	Latex House Paint	Water Repellent
Wood Surfaces												
Clapboard	X.			X					X		X.	
Natural Wood Siding & Trim			X		X							
Shutters & Other Trim	X.						X.		X		X.	
Wood Frame Windows	X.			X			X.		X		X.	
Wood Porch Floor								X				
Wood Shingle Roof					X							X
Metal Surfaces												
Aluminum Windows	X.			X			X.			X	X.	
Steel Windows	X.			X.			X.			X	X.	
Metal Roof	X.									X	X.	
Metal Siding	X.			X.			X.			X	X.	
Copper Surfaces			X									
Galvanized Surfaces	X.			X.			X.			X	X.	
Iron Surfaces	X.			X.			X.			X	X.	
Miscellaneous												
Asbestos Cement	X.								X		X	
Brick	X.	X		X					X		X	X
Cement & Cinder Block	X.	X		X					X		X	
Concrete/Masonry Porches And Floors								X			X	
Coal Tar Felt Roof						X						
Stucco X.	X.	X		X					X		X	

• dot at right of X indicates a primer or sealer may be needed before finishing coat is applied

SOURCE: U.S. Department of Commerce

Notes: ________________

ROOM ARRANGEMENTS-----

THE FOLLOWING PAGES HAVE SKETCHES, INFORMATION AND DATA THAT WILL ASSIST YOU IN DESIGNING AND PLANNING ROOMS FOR RESIDENTIAL USE. WHETHER FOR A NEW HOME, RENOVATION, REMODELING OR RESTORATION PROJECT, YOU CAN COORDINATE YOUR EFFORTS AND THINKING BY USING THIS INFORMATION.

FURNITURE PLACEMENT, LOCATION OF DOORS, WINDOWS AND CLOSETS ARE SUGGESTED. NORMAL REQUIRED AND PRACTICAL CLEARANCES ARE SHOWN TO GUIDE YOU IN PROVIDING FUNCTIONAL, COMFORTABLE ROOMS FOR YOUR HOME. BEDROOMS, DINING AND LIVING AREAS ARE ALSO COVERED. TO GIVE YOU COMPLETE DATA AND SUGGESTIONS ON RESIDENTIAL INTERIOR DESIGN AND PLANNING, YOU WILL FIND KITCHEN AND BATHROOM DETAILS IN OTHER SECTIONS OF THIS BOOK.

REFER TO THE INDEX FOR COMPLETE INFORMATION ON THESE SUBJECTS. USE THE SPACES PROVIDED THROUGHOUT THIS BOOK TO TAKE NOTES, MEASUREMENTS AND OTHER INFORMATION—AND SAVE TIME AS YOU DEVELOP AND PLAN YOUR HOME IDEAS.

THE FURNITURE TYPES AND SIZES USED IN THESE ARRANGEMENTS ARE SHOWN IN DETAIL ON THE PAGE ENTITLED *ROOM ARRANGEMENTS, FURNITURE TYPES, SIZES*.

ROOM ARRANGEMENTS, BEDROOMS-----

ADEQUATE CLOSETS AND STORAGE AREAS SHOULD BE PROVIDED IN ALL LIVING AREAS OF THE HOME. IN THE INTEREST OF SPACE SAVING AND ECONOMY THEY OFTEN WILL ACQUIRE ODD SHAPES AND SIZES. FOR THIS REASON, SIZES, TYPES, FLOOR PLANS, SHAPES AND DIMENSIONS ARE NOT RECOMMENDED HERE.

THE DIMENSIONS, SIZES AND SPACE REQUIREMENTS FOR CLOTHING AND OTHER TYPES OF WEARING APPAREL NORMALLY STORED IN CLOSETS ARE SHOWN UNDER *APPAREL, SPACE REQUIRED* IN THIS BOOK.

THE AVERAGE HOME TODAY HAS TWO OR MORE BEDROOMS AND OFTEN ADDITIONAL SLEEPING ARRANGEMENTS ARE PROVIDED IN OTHER AREAS.

MOST OF THE BEDROOMS IN A HOME ARE USUALLY DIFFERENT SIZES. THIS SECTION SHOWS SEVERAL BEDROOM ARRANGEMENTS, INCLUDING ONE WITH BABY CRIB SPACE INDICATED.

THE LOCATION OF DOORS AND WINDOWS SHOULD PERMIT ALTERNATE FURNITURE ARRANGEMENTS.

TO ASSURE ADEQUATE SPACE FOR CONVENIENT USE OF FURNITURE IN THE BEDROOM, NOT LESS THAN THE FOLLOWING CLEARANCES SHOULD BE OBSERVED.

42" AT ONE SIDE OR FOOT OF BED FOR DRESSING
12" BETWEEN SIDE OF BED AND SIDE OF DRESSER OR CHEST
36" IN FRONT OF DRESSER, CLOSET AND CHEST OF DRAWERS
24" FOR MAJOR CIRCULATION PATH (DOOR TO CLOSET, ETC.)
22" ON ONE SIDE OF BED FOR CIRCULATION
12" MINIMUM ON LEAST USED SIDE OF BEDS

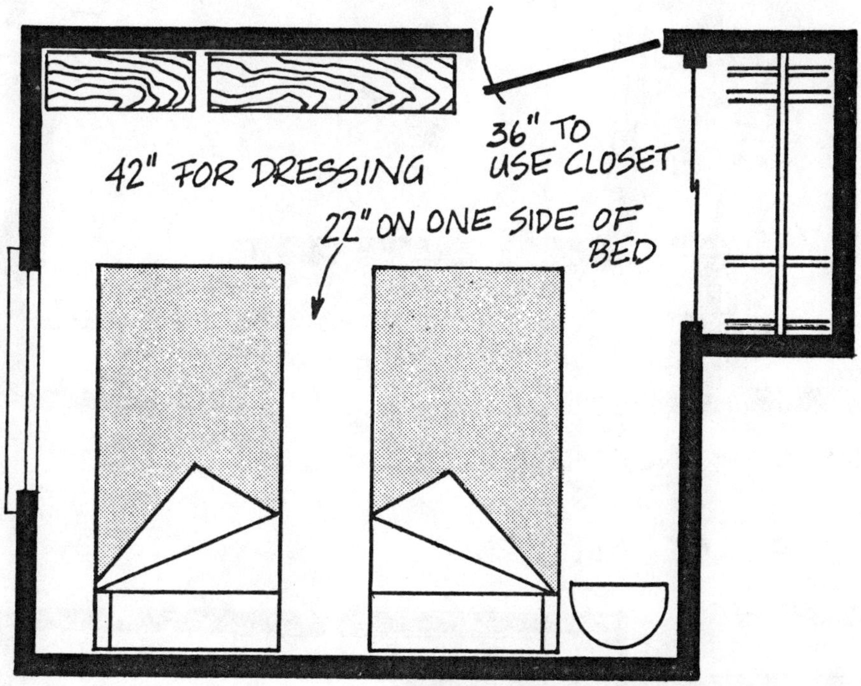

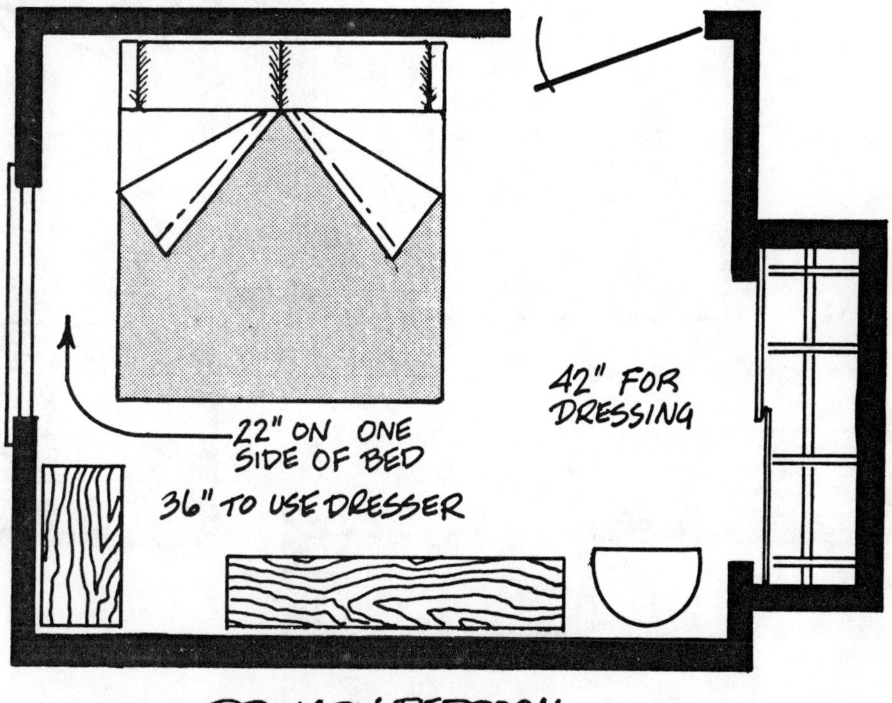

PRIMARY BEDROOM

ROOM ARRANGEMENTS, BEDROOMS (continued)

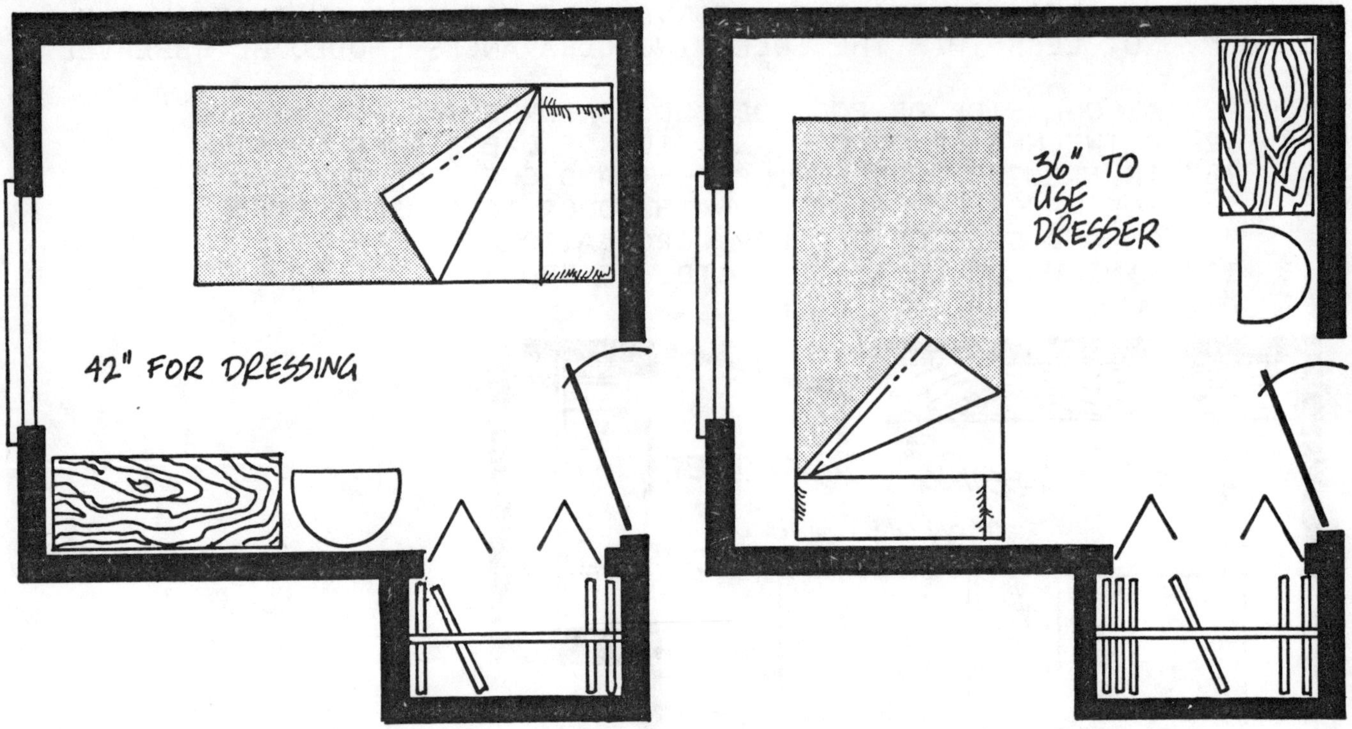

SINGLE OCCUPANCY BEDROOM

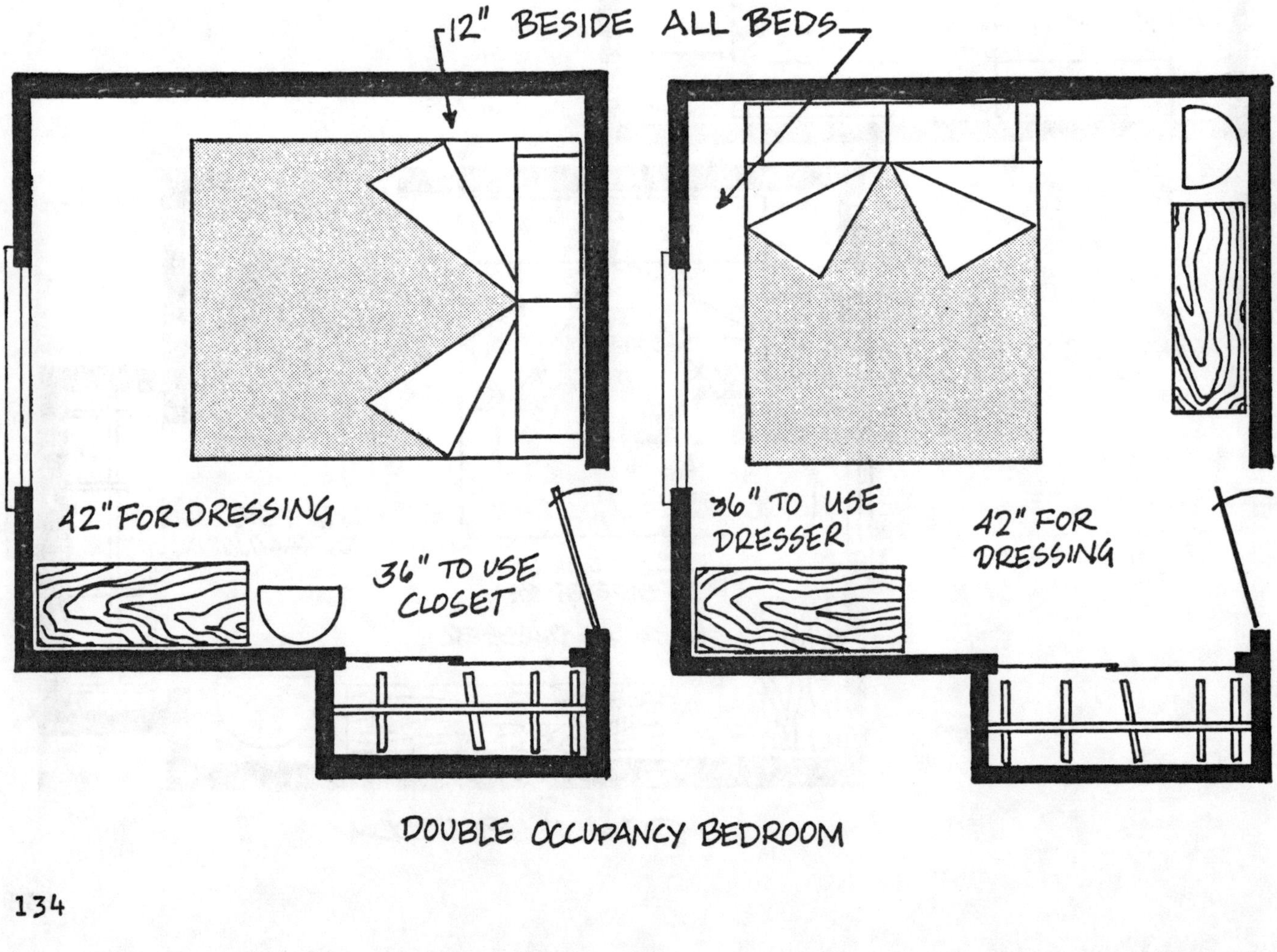

DOUBLE OCCUPANCY BEDROOM

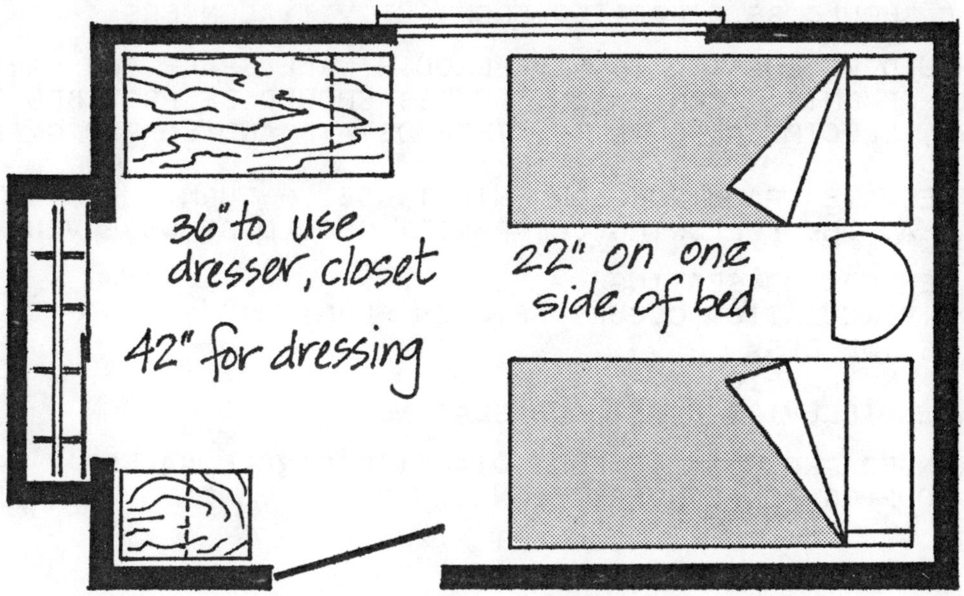

ROOM ARRANGEMENTS, DINING AREAS-----

TO ASSURE ADEQUATE SPACE FOR CONVENIENT USE OF THE DINING AREA, THE FOLLOWING CLEARANCES FROM THE EDGE OF THE DINING TABLE SHOULD BE OBSERVED:

- 32" FOR CHAIRS PLUS ACCESS THERETO
- 38" FOR CHAIRS PLUS ACCESS AND PASSAGE
- 42" FOR SERVING FROM BEHIND CHAIR
- 24" FOR PASSAGE ONLY
- 48" FROM TABLE TO BASE CABINET (IN DINING-KITCHEN)

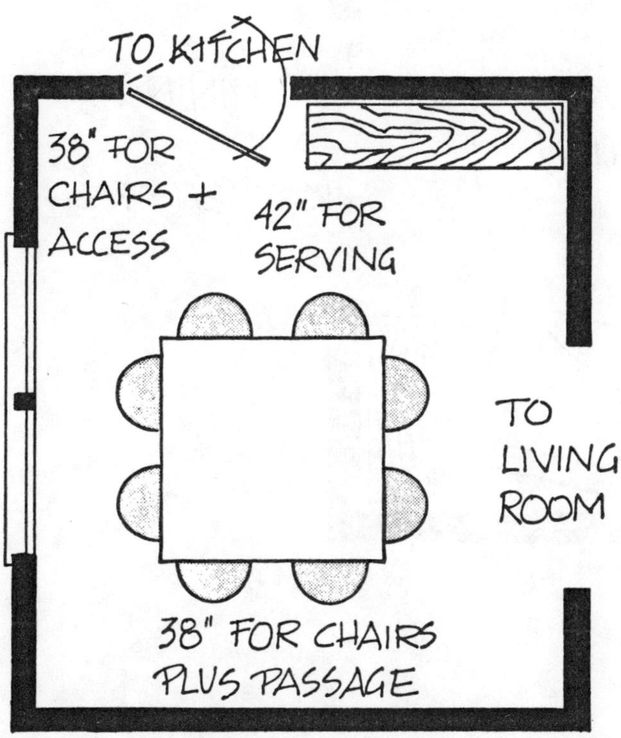

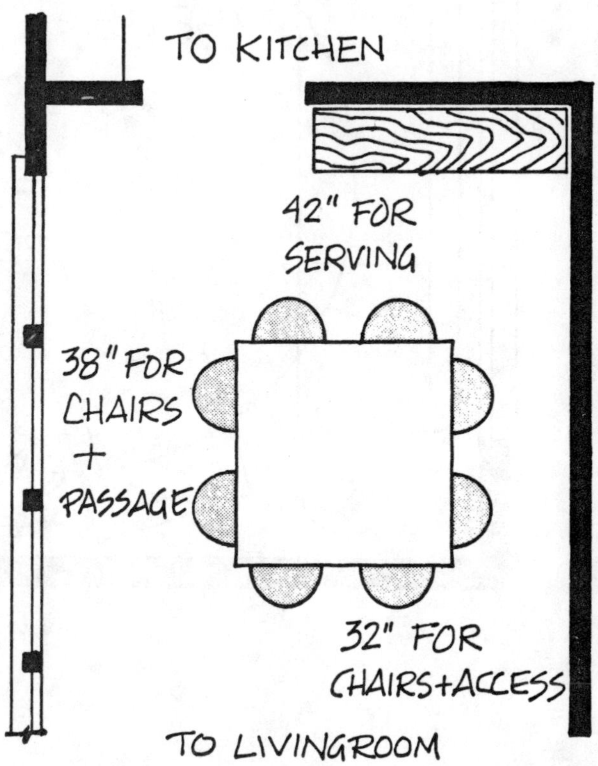

ROOM ARRANGEMENTS, LIVING ROOM-----

THRU TRAFFIC SHOULD BE SEPARATED FROM ACTIVITY CENTERS.

OPENINGS SHOULD BE LOCATED TO GIVE ENOUGH WALL SPACE FOR VARIOUS FURNITURE ARRANGEMENTS. CONVENIENT ACCESS SHOULD BE PROVIDED TO DOORS, WINDOWS, ELECTRICAL OUTLETS, THERMOSTATS AND SUPPLY GRILLS.

TO ASSURE ADEQUATE SPACE FOR CONVENIENT USE OF FURNITURE IN THE LIVING ROOM AREA, THE FOLLOWING CLEARANCES SHOULD BE OBSERVED:

60" BETWEEN FACING SEATING
24" WHERE CIRCULATION OCCURS BETWEEN FURNITURE
30" FOR USE OF DESK
36" FOR MAIN TRAFFIC
60" BETWEEN TELEVISION SET AND SEATING

SEATING ARRANGED AROUND A 10 FOOT DIAMETER CIRCLE MAKES A COMFORTABLE GROUPING FOR CONVERSATION.

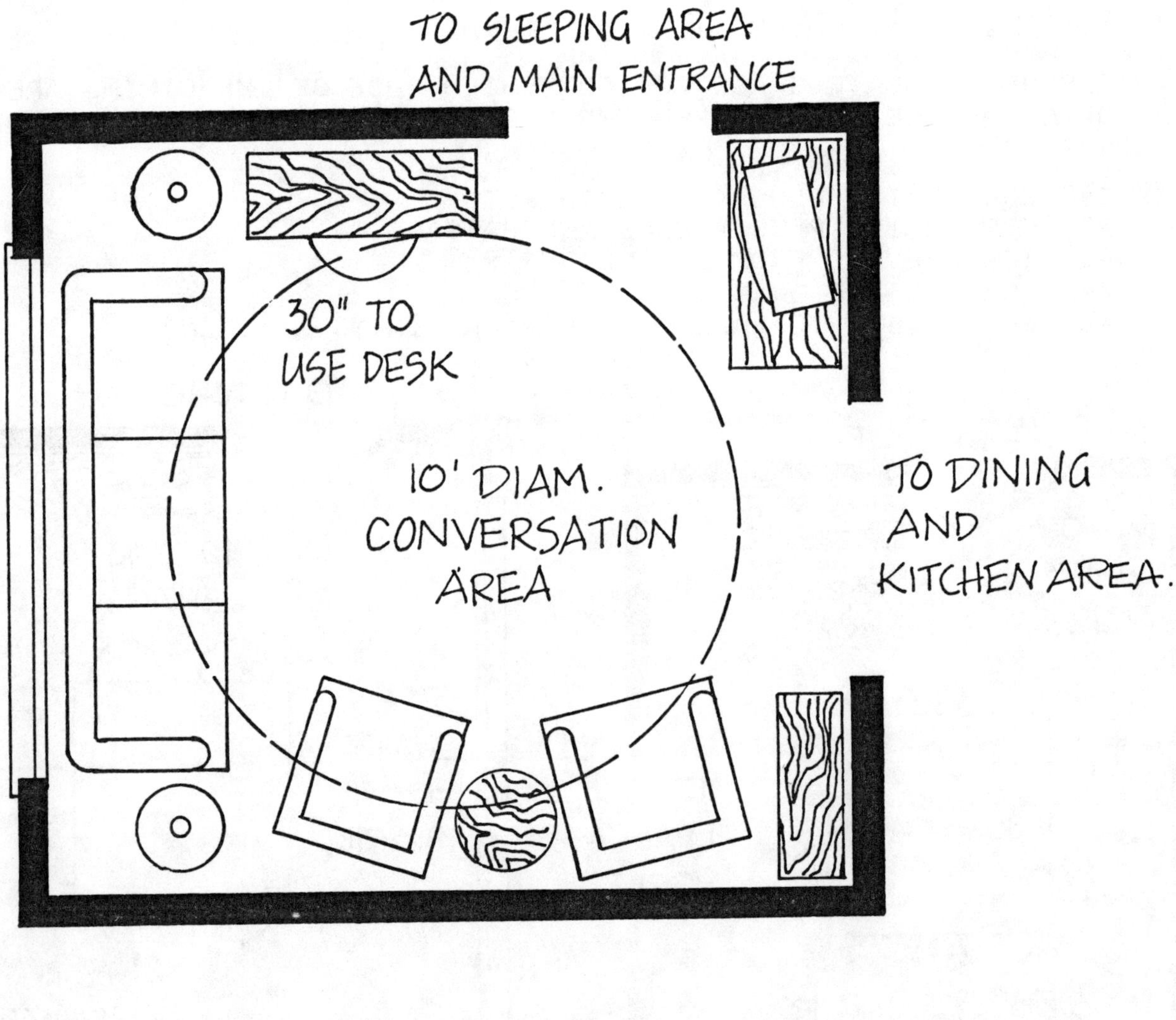

ROOM ARRANGEMENTS, FURNITURE TYPES, SIZES-----

THE FURNITURE TYPES, SIZES AND DATA INDICATED ON THIS PAGE WILL HELP YOU COORDINATE YOUR DESIGN AND PLAN WITH THE ROOM ARRANGEMENTS SHOWN ON THE PRECEDING PAGES. THIS INFORMATION WILL PROVE HELPFUL IN FURNITURE SELECTION, PLACEMENT, CLEARANCES, DESIGN AND TYPE.

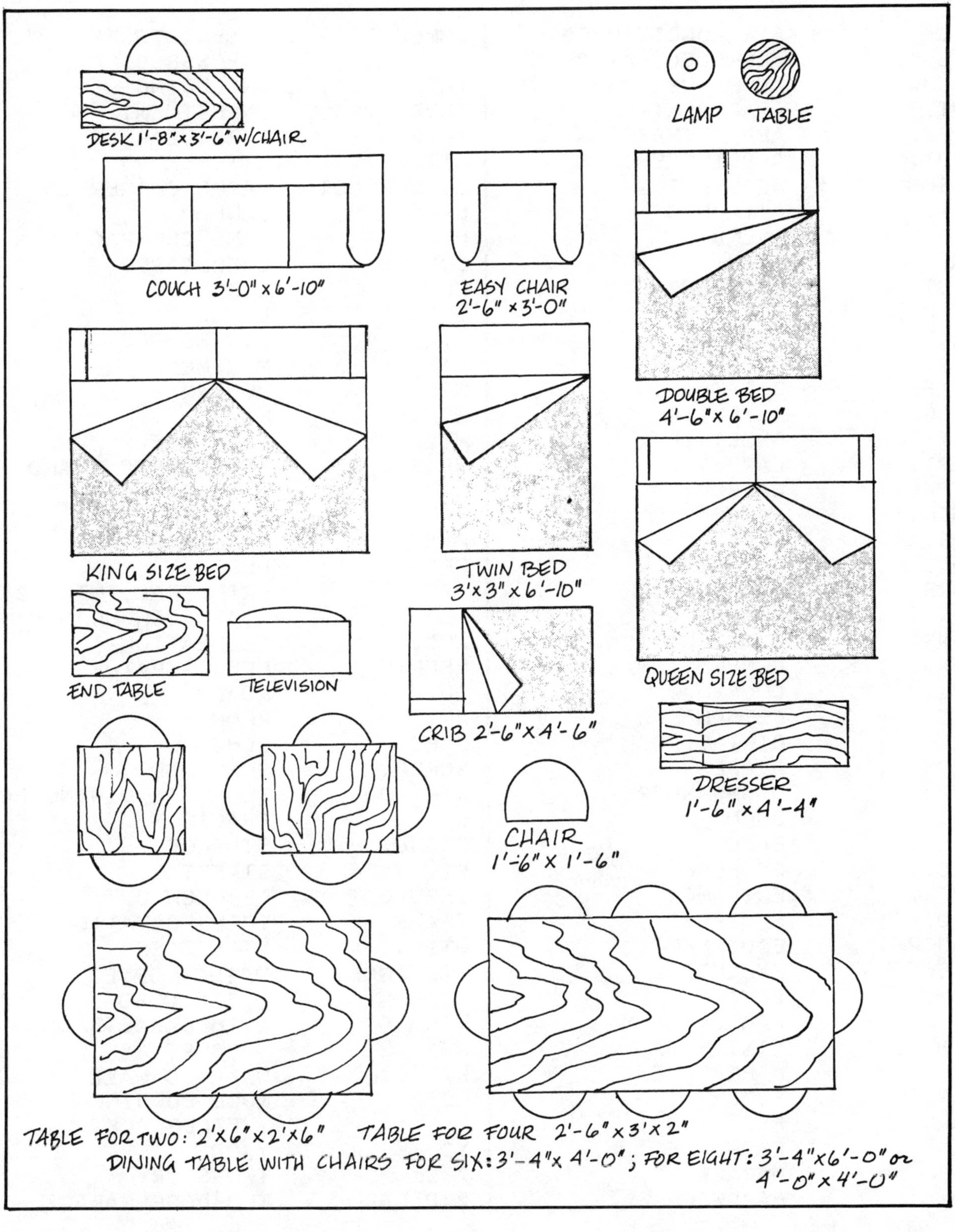

REAL ESTATE TERMS AND ABBREVIATIONS-----

IN THE PURCHASE OR SALE OF REAL PROPERTY, MANY TERMS AND ABBREVIATIONS ARE USED BY THE TRADE IN AGREEMENTS, SALES CONTRACTS, LEASES, DEEDS, ADVERTISEMENTS AND DESCRIPTION OF PROPERTY. THIS LIST WILL BE HELPFUL TO YOU IN ANY TRANSACTION.

A/C	AIR CONDITIONED
A.E.K.	ALL-ELEC. KITCHEN
ALUM.	ALUMINUM
APT.	APARTMENT
APPX.	APPROXIMATELY
APPT.	APPOINTMENT
ASSUM.	ASSUMPTION
ATT.	ATTACHED
AVE.	AVENUE
BA	BATH
BSMT.	BASEMENT
BEAUT.	BEAUTIFUL
BR	BEDROOM
BLK.	BLOCK
BRKFST	BREAKFAST
BLDG.	BUILDING
BLT-IN	BUILT-IN
CRPT.	CARPET
C/P	CARPORT
CATH.	CATHEDRAL
CER.	CERAMIC
COL.	COLONIAL
COMB.	COMBINATION
CONDT.	CONDITION
CONST.	CONSTRUCTION
CONV.	CONVENTIONAL
CT.	COURT
DET.	DETACHED
DR	DINING ROOM
DRPS.	DRAPES
D.W.	DISHWASHER
DN.	DOWN
ELEC.	ELECTRIC
ENCL.	ENCLOSED
ENTR.	ENTRANCE
EQUIPT.	EQUIPPED
EQUIPMT.	EQUIPMENT
EXEC.	EXECUTIVE
FAM.	FAMILY
FR	FAMILY ROOM
FIN.	FINISHED
FRPL.	FIREPLACE
FLR.	FLOOR
FLRG.	FLOORING
FRNTG.	FRONTAGE
GAR.	GARAGE
HDWD.	HARDWOOD
IMMED.POSS.	IMMEDIATE POSSESSION
INCL.	INCLUDING
INVEST.	INVESTMENT
IRREG.	IRREGULAR
KIT.	KITCHEN
LG COR LOT	LARGE CORNER LOT
LN.	LANE
L.C.	LAND CONTRACT
LDSCPD	LANDSCAPED
LGE.	LARGE
LVL.	LEVEL
L.R.	LIVING ROOM
MIN.	MINIMUM
MOD.	MODERN
MON.	MONTH
MTGE.	MORTGAGE
NR S & S	NEAR SCHOOLS AND SHOPPING
PAN.	PANELED
PYMTS.	PAYMENTS
PRCH.	PORCH
PITI	PRINCIPAL, INTEREST, TAX & INSURANCE
REC.	RECREATION
REFRIG.	REFRIGERATOR
RD.	ROAD
RM.	ROOM
SCHL.	SCHOOL
SCRND.	SCREENED
SEP. DR	SEPARATE DINING ROOM
SPAC.	SPACIOUS
STRGE.	STORAGE
ST.	STREET
THRU-OUT	THROUGH OUT
TRANS.	TRANSPORTATION
UTIL.	UTILITIES
W/D INCL.	WASHER/DRYER INCLUDED
W BB CT.	WITH BASKETBALL COURT
W/W	WALL TO WALL
WBFP	WOOD BURNING FIREPLACE
YD.	YARD
2-S	TWO-STORY
2/D GAR	TWO-DOOR GARAGE

RESTORATION-----

RESTORATION OF OLDER HOMES TAKES ON NEW MEANING EVERY DAY AS THE LIFESTYLE OF OUR COUNTRY CHANGES. AS MANY GOOD, WELL-BUILT, STRATEGICALLY LOCATED HOMES BECOME OLDER AND OBSOLETE, THEY CAN BECOME MORE DESIRABLE—AND MORE ADAPTABLE—TO COMPLETE RESTORATION. GOOD BUILDING SITES CONVENIENTLY LOCATED IN DESIRABLE AREAS AND NEIGHBORHOODS ARE LIMITED AND VERY EXPENSIVE.

THIS, COUPLED WITH THE CONTINUED INFLATION FACTOR AND THE RISING COSTS OF BUILDING, HAS DEVELOPED AN ENTIRELY NEW FIELD AND DESIRE FOR THE RENOVATION OF THE OLDER HOME.

WHILE THE MAIN STRUCTURE AND OTHER COMPONENT PARTS OF A GOOD HOME HAVE AN UNLIMITED LIFE SPAN IF PROPERLY MAINTAINED, MANY PARTS HAVE A LIMITED LIFE AND IF WORN OUT MUST BE REPLACED OR UPDATED TO MEET NEW CODES AND REGULATIONS. THIS COVERS ITEMS SUCH AS HEATING, WIRING, PLUMBING, ROOFING, WINDOWS, DOORS, FLOORS, BATHS, KITCHENS, ETC. IF YOU CONTEMPLATE A FULL RESTORATION PROJECT YOU MAY DESIRE TO MODERNIZE MANY ITEMS WHILE IN THE PROCESS—EVEN THOUGH SOME OF THEM ARE STILL SERVICEABLE.

FOLLOWING IS A LIST OF A FEW OF THE MANY ITEMS YOU SHOULD CONSIDER AS YOU PLAN THE RESTORATION OF AN OLDER HOME. IF DESIRED, YOU CAN SECURE ASSISTANCE THROUGH A CONTRACTOR OR LENDING INSTITUTION OF YOUR CHOICE BEFORE YOU PROCEED WITH PLANNING OR INITIAL WORK.

AS YOU USE THIS BOOK YOU WILL FIND A GREAT MANY THINGS THAT WILL PROVE VERY HELPFUL TO YOU IN CONSIDERING AND PLANNING YOUR RESTORATION PROJECT. CONSULT AND USE THE COMPLETE INDEX TO FIND THE ITEMS YOU WANT TO CHECK.

SEE: FINDING & FIXING THE OLDER HOME (STRUCTURES PUBLISHING CO.)

Notes:

RESTORATION CHECK LIST-----

AGE OF HOUSE______ ADAPTABILITY TO NEEDS______________________

BASEMENT______ CONDITION______ FLOOR______ WALLS______ BATH______

CHIMNEYS__________ FLUES__________ CEILINGS____________________

DOORS (EXIT)__________(INTERIOR)__________ DRIVES______________

ELECTRICAL EQUIPMENT____________________ FIREPLACE_____________

FLOORS__________ FOUNDATION__________ FLASHING, ROOF____________

GARAGE____________ HARDWARE____________ HEATING________________

ADAPT TO AIR CONDITIONING______________________________________

INSULATION____

INTERIOR WALLS_________________ EXTERIOR WALLS________________

LIGHTING_______________ LOCATION__________ MASONRY WORK__________

PLASTER__________ PAINT (INTERIOR)__________(EXTERIOR)__________

PLUMBING________________________ PLUMBING FIXTURES_______________

PORCHES____________ PATIOS____________ ROOF STRUCTURE____________

ROOFING_______________ GUTTERS_______________ EAVES______________

DOWN SPOUTS__________ SEWERS__________ SIDING____________________

STAIRS__________________ STRUCTURAL CONDITION____________________

FLOOR JOIST__________ STUDS__________ TERMITE CHECK______________

WINDOWS__________________________.

OTHER ____

____

____

____

____

____

____

____

____

____

____

____

ROOFING-----

ASPHALT STRIP SHINGLES ARE USED MUCH MORE WIDELY ON ALL TYPES OF HOUSING TODAY THAN ALL OTHER ROOFING MATERIALS COMBINED. THEY NORMALLY ARE APPLIED OVER A LAYER OF 15 POUND ROOFING FELT AND ARE NAILED IN PLACE WITH LARGE-HEADED CORROSION RESISTANT ROOFING NAILS.

THESE 1' X 3' STRIP SHINGLES MAY BE REFERRED TO AS "3 TAB" SHINGLES BECAUSE THE COMMON TYPE HAS TWO 1/2 INCH CUT OUTS AT ONE FOOT INTERVALS TO GIVE IT THREE SHINGLE TABS. HOWEVER, IN RECENT YEARS SOME SHINGLE MANUFACTURERS ARE OFFERING TABLESS TYPES AND ALSO ARE PRODUCING IRREGULAR BUTT AND DOUBLE LAYER LAMINATED TYPES FOR AN IMPROVED TEXTURED APPEARANCE. THE SIZE OF THESE REMAINS 1' X 3'.

THE TREND HAS BEEN TOWARD HEAVIER SHINGLES WITH MORE COLORS AND STYLES. FOR MANY YEARS, THE STANDARD, WIDELY USED, WEIGHT OF ASPHALT SHINGLE COVERAGE WAS IN THE RANGE OF 210 TO 240 POUNDS PER 100 SQUARE FEET. TODAY, WITH NEWER MANUFACTURING FACILITIES, DOUBLE LAYERING AND THE USE OF LARGER, HEAVIER MINERAL GRANULES, SHINGLE WEIGHTS RUN TO 300 POUNDS OR MORE PER SQUARE (100 SQUARE FEET).

A FEW YEARS AGO, THE FIBERGLASS TYPE OF ASPHALT SHINGLES WERE INTRODUCED AND ARE BECOMING MORE WIDELY USED BECAUSE OF THEIR SUPERIOR QUALITIES. IN THESE NEWER SHINGLES, THE TRADITIONAL FELT BASE MATERIAL IS REPLACED BY A GLASS FIBER MAT WHICH IS STRONGER AND LESS BULKY THAN THE FELT. AS A RESULT, FIBERGLASS SINGLES CONTAIN ABOUT 50 PERCENT MORE ASPHALT FOR THE SAME OVER-ALL WEIGHT OF SHINGLE. THIS MEANS BETTER WEATHER RESISTANCE AND LONGER LIFE. ALSO, THE FIBERGLASS SHINGLES ARE MORE FIRE RESISTANT AND HAVE EARNED A "CLASS A" FIRE RATING FROM UNDER-WRITERS' LABORATORIES. THIS IS IN CONTRAST TO THE "CLASS C" RATING WHICH IS THE HIGHEST THAT FELT BASE SHINGLES CAN EARN.

VIRTUALLY ALL ASPHALT STRIP SHINGLES TODAY HAVE A SERIES OF ASPHALT-ADHESIVE SPOTS STRATEGICALLY LOCATED TO PROVIDE A SELF SEAL-DOWN ACTION TO GIVE MORE EFFECTIVE WIND RESISTANCE.

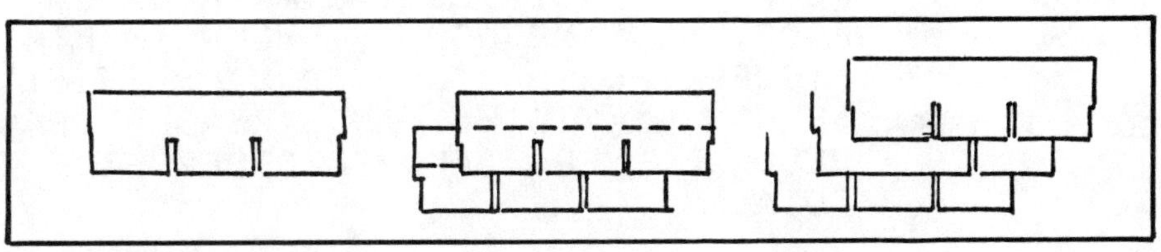

ROOFING (continued)

ROLL ROOFING IS NOT AS READILY AVAILABLE AS IT ONCE WAS. IT NO LONGER IS COMMONLY USED ON HOMES. IN RURAL AREAS, IT STILL SEES USE ON SHEDS AND OUTBUILDINGS. IT COMES WITH LIMITED COLOR CHOICE IN BLACK, RED, GREEN AND BLUE MINERAL SURFACED COLORS. ROLLS ARE 36" WIDE AND CONTAIN 108 SQUARE FEET.

ROOF AREAS-----

IF YOU WANT TO DETERMINE THE AREA OF A ROOF OR WANT TO CHECK THE QUANTITIES OF MATERIALS REQUIRED FOR A ROOF, YOU WILL FIND THESE SKETCHES AND ACCOMPANYING INFORMATION VERY HELPFUL AND EASY TO FOLLOW.

PLAIN GABLE ROOF, FIG, 1

MULTIPLY THE LENGTH OF THE ROOF B-C BY THE RAFTER LENGTH A-B. THIS GIVES THE AREA OF ONE SIDE. DOUBLE THIS TO GET THE ENTIRE AREA.

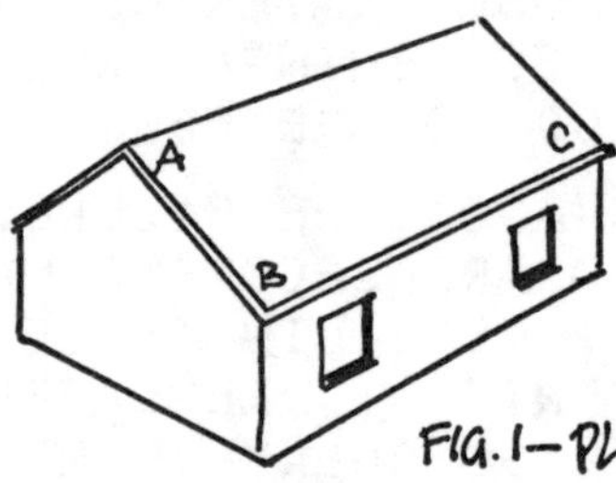

FIG. 1— PLAIN GABLE ROOF

PLAIN GABLE ROOF WITH DORMERS, FIG. 2

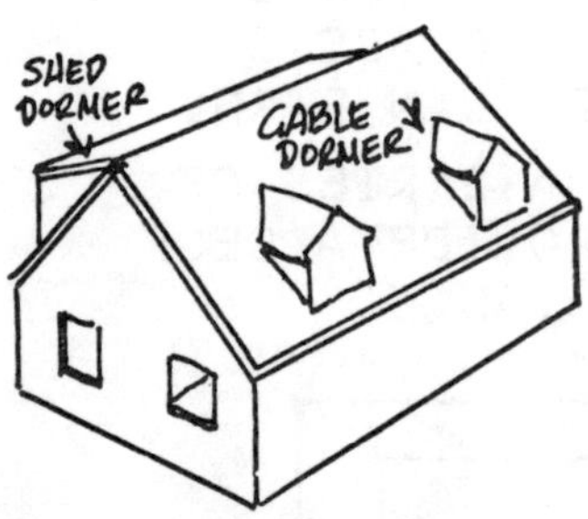

FIG. 2- GABLE ROOF WITH DORMER

FIGURE EACH SIDE OF THE ROOF SEPARATELY. USE THE ABOVE METHOD TO GET THE AREA OF THE RIGHT SIDE WITH THE TWO DORMERS AS SHOWN. DEDUCT THE AREAS WHERE THE DORMERS INTERRUPT THE ROOF SURFACE. ADD THE DORMER ROOFS BY FIGURING THE AREA OF EACH DORMER ROOF. USE METHOD AS SHOWN IN FIG. 3 BELOW. FIGURE THE SHED DORMER BY MULTIPLYING LENGTH BY WIDTH. ADD AREAS AT EACH END AND BELOW DORMER. ADD TO OTHER SIDE TO GET TOTAL.

HIP ROOF, FIG. 3

MULTIPLY LENGTH OF EAVE A-C BY 1/2 THE LENGTH B-D. THIS IS THE AREA OF ONE END. MULTIPLY BY 2 TO GET BOTH ENDS. ADD RIDGE LENGTH B-E TO EAVE LENGTH C-F. DIVIDE THIS BY 2 AND MULTIPLY BY BY RAFTER LENGTH E-G. MULTIPLY BY 2 AND ADD TO AREA OF ENDS TO GET TOTAL ROOF AREA.

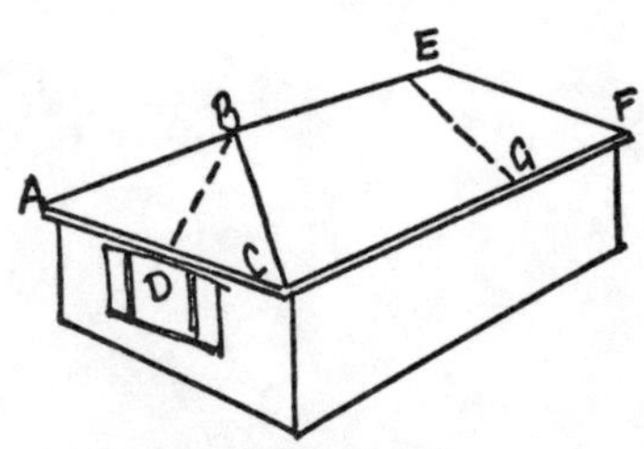

FIG. 3- HIP ROOF

ROOF AREAS *(continued)*

FIG. 4—GAMBREL ROOF

GAMBREL ROOF, FIG. 4

MULTIPLY LENGTH OF ROOF A-B BY TOTAL WIDTHS OF SECTIONS A-C AND C-D. THIS GIVES ONE HALF OF ROOF. MULTIPLY BY 2 TO GET TOTAL AREA.

FLAT ROOF, NOT SHOWN.

MULTIPLY LENGTH BY WIDTH. ADD 5 PERCENT TO 20 PERCENT AREAS DEPENDING UPON ROOF TYPE TO COVER CUTTING, WASTE, RIDGES, HIPS, ETC.

NOTE: KEEP LENGTHS OF RIDGES, VALLEYS, EAVES AND DORMER SIDES SEPARATE FOR CONVENIENCE WHEN FIGURING THESE ITEMS.

Notes:

ROOF PITCH-----

THE PITCH OF THE ROOF DETERMINES THE KIND AND TYPE OF MATERIALS YOU SHOULD USE ON YOUR HOUSE. THIS RANGES FROM BUILT-UP TYPE ON ALMOST FLAT OR LOW PITCHED ROOF AREAS TO ALL TYPES AND STYLES OF SHINGLES AND TILE ON HIGH PITCHED ROOFS.

ASPHALT OR FIBRE GLASS FLEXIBLE, SELF-SEALING SHINGLES MAY BE USED ON AS LOW AS 2" PITCHED ROOFS. NORMAL SHINGLES OF THIS TYPE AND APPLIED IN THE USUAL MANNER WILL USUALLY REQUIRE 4" PITCH. THE DIRECTIONS OF THE MANUFACTURERS SHOULD ALWAYS BE FOLLOWED WHEN LAYING ROOFING MATERIALS TO INSURE A GOOD JOB.

RISE, RUN AND PITCH-----

RISE IS THE DISTANCE BETWEEN THE EAVE LEVEL AND RIDGE.

RUN IS THE LENGTH OF THE RAFTERS PROJECTED TO THE HORIZONTAL.

PITCH IS THE RAFTER ANGLE. THESE TERMS ARE SHOWN IN THE ACCOMPANYING SKETCH AT THE RIGHT.

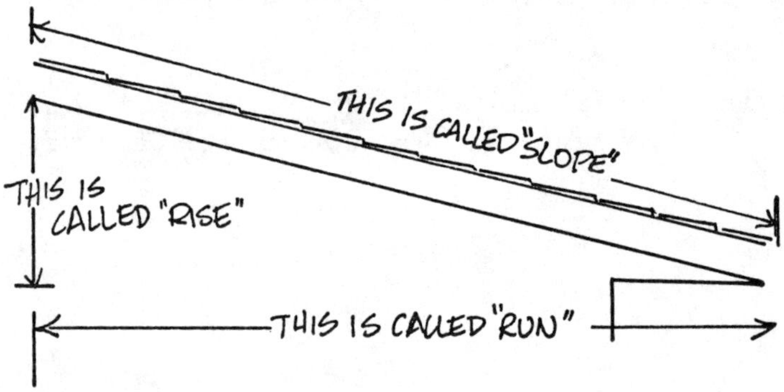

1/4TH PITCH IS 6" RISE TO 12" RUN.

1/3RD PITCH IS 8" RISE TO 12" RUN.

1/2 PITCH IS 12" RISE TO 12" RUN.

FOR EXAMPLE, IF YOU HAVE A HOUSE 24' WIDE AND THE RIDGE IS 12' ABOVE THE WALL LINE, IT WOULD BE 1/2 PITCH OR 24' DIVIDED BY 12' EQUALS 1/2. A QUITE COMMON WAY OF REFERRING TO ROOF SLOPE, RATHER THAN BY FRACTIONS, IS BY THE RELATION BETWEEN RISE AND RUN: 6 IN 12, 8 IN 12, 12 IN 12 - ALSO: 2-1/2 IN 12 OR 9-1/4 IN 12.

Notes:

ROOF SLOPE AND PITCH-----

THE SLOPE OR PITCH OF A ROOF REQUIRES CERTAIN MATERIALS AND INSTALLATION PROCEDURES BE USED TO INSURE PROPER AND SAFE WEATHER TIGHTNESS AND TO CONFORM TO LOCAL CODE REQUIREMENTS AND REGULATIONS.

ROOF SLOPE OR PITCH IS COMMONLY REFERRED TO IN RUN OR RISE NUMBERS; THAT IS 4 INCHES IN 1 FOOT.

THE CHART BELOW DEALS WITH COMPOSITION ROLL ROOFING AND SHINGLES OF ALL TYPES, AS SHOWN IN *ROOFING* IN THIS BOOK.

THE SUGGESTED USE OF ROLL ROOFING AND ASPHALT OR FIBRE GLASS SHINGLES AND THE METHOD OF APPLICATION IS CLEARLY INDICATED FOR THE VARIOUS DEGREES OF ROOF PITCH.

THE CHART BELOW DEALS WITH COMPOSITION ROLL ROOFING AND SHINGLES OF ALL TYPES, AS SHOWN IN *ROOFING* IN THIS BOOK.

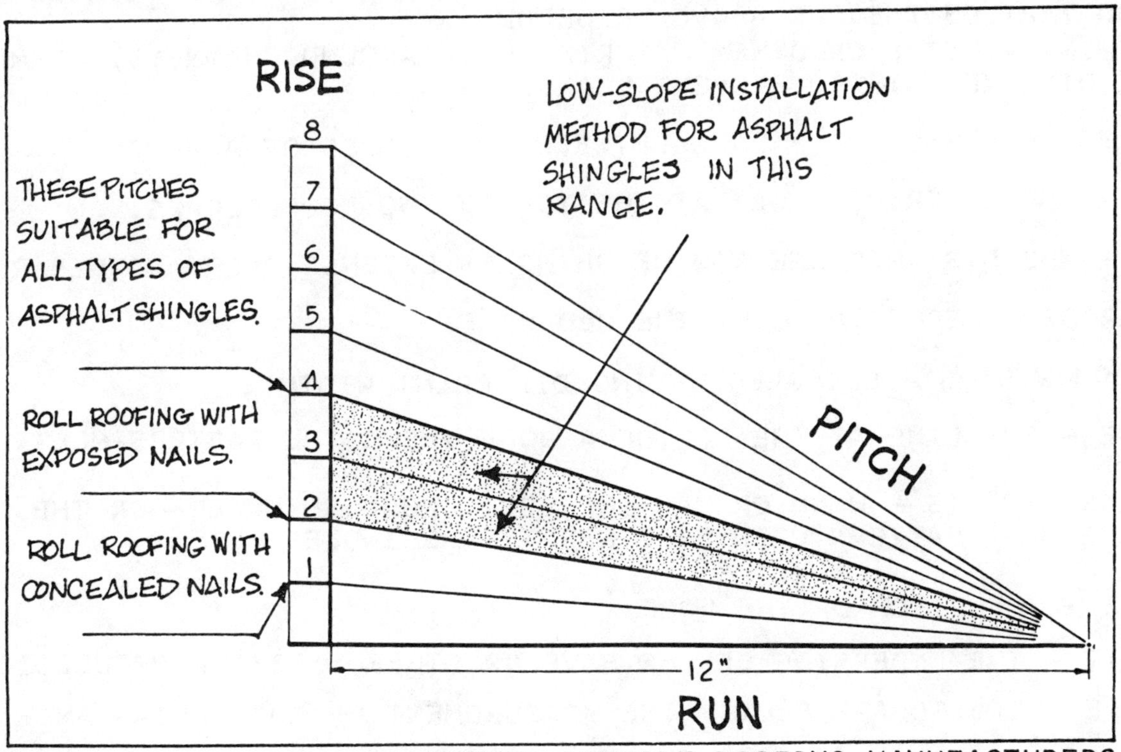

COURTESY ASPHALT ROOFING MANUFACTURERS

SEE: *ROOF: PITCH, RUN, RISE*

Notes:

ROOFING TERMS-----

AS YOU WILL FIND IN TALKING ABOUT HOUSES, MOST PARTS HAVE CERTAIN TERMS WHICH ARE COMMONLY USED BY THE TRADE. HERE ARE SOME OF THE TERMS AND DEFINITIONS RELATED TO ROOFING.

- BUILT-UP - ROOFING FOR FLAT OR NEAR FLAT AREAS. IT IS COMPOSED OF LAYERS OF FELT CEMENTED TOGETHER BY HOT TAR OR ASPHALT.
- BUTT END - THE EXPOSED ENDS OF SHINGLES AS YOU LOOK AT THE ROOF.
- CANVAS DECK - CANVAS TOPPED ROOFING FOR AREAS THAT REQUIRE WEAR BY FOOT TRAFFIC OR SERVICE.
- DECKING - ANOTHER NAME FOR ROOF SHEATHING.
- DORMER - A FRAME SECTION OF WINDOW PROJECTING FROM A ROOF SLOPE.
- EXPOSURE - MEASURE IN INCHES BETWEEN THE BUTT OF ONE SHINGLE AND THAT OF THE ONE ABOVE OR BELOW.
- FLASHING - METAL OR OTHER MATERIAL USED AROUND CHIMNEYS, DORMERS OR OTHER POINTS TO PREVENT LEAKS.
- GABLE - THE PART OF A WALL BETWEEN THE SLOPES OF A ROOF.
- HIP - THE EXTERIOR ANGLE AT THE TOP OF TWO ROOF SLOPES.
- LAP - THE DISTANCE ONE ROW OF SHINGLES EXTENDS OVER ANOTHER ROW.
- OVERROOF - REROOFING OVER THE OLD ROOF.
- RANDOM WIDTHS - SHINGLES HAVING DIFFERENT WIDTHS.
- RIDGE - THE LINE AT THE TOP OF A ROOF WHERE THE RAFTERS MEET.
- SHAKES, SPLITS - HAND OR MACHINE SPLIT WOOD SHINGLES—OR THE NAME OF A PATTERN USED BY OTHER TYPE SHINGLES.
- SLOPE - THE PITCH OF THE ROOF.
- SOFFIT - THE UNDERSIDE OF THE EAVE OR OTHER EXTERIOR PROJECTION.
- SQUARE - 100 SQUARE FEET. THE MEASUREMENT OF ROOF AREAS AND MATERIALS.
- STAGGERED BUTTS - SHINGLES LAID WITH AN UNEVEN LAP TO GIVE THE ROOF TEXTURE AND PATTERN.
- STARTERS - THE FIRST COURSE OF SHINGLES OR TILE ROOFING.
- THICK BUTTS - TAPERED SHINGLES THAT ARE THICK AT THE EXPOSED ENDS.
- VALLEY - WHERE TWO SLOPES OF ROOF MEET TO FORM A V DESIGN.

SEE: *PITCH, ROOF: PITCH, RUN, RISE.*

ROOF DESIGNS-----

THE STYLE AND DESIGN OF THE ROOF STRUCTURE YOU CONSIDER FOR YOUR NEW HOME, REMODELING OR RESTORATION PROJECT WILL DEPEND ON THE TYPE BUILDING, ATTIC, STORAGE AREAS, ETC., YOU WILL REQUIRE. THE TYPES AND DESIGNS COMMONLY USED ARE SHOWN HERE. VARIATIONS, ADJUSTMENTS AND COMBINATIONS MAY BE DEVELOPED TO SUIT YOUR DESIRES.

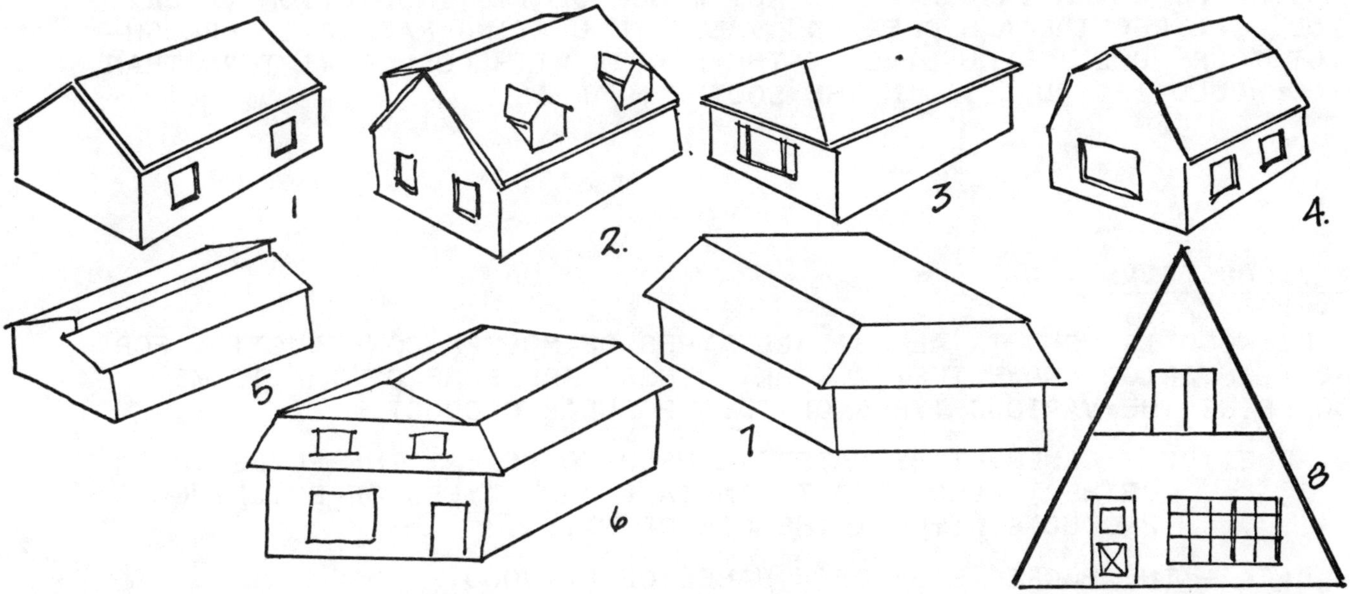

1. GABLE ROOF
2. GABLE ROOF W/ DORMER
3. HIP ROOF
4. GAMBREL ROOF
5. SHED ROOF
6. MANSARD ROOF
7. MANSARD EAVE TYPE
8. A FRAME

ROOF PARTS-----

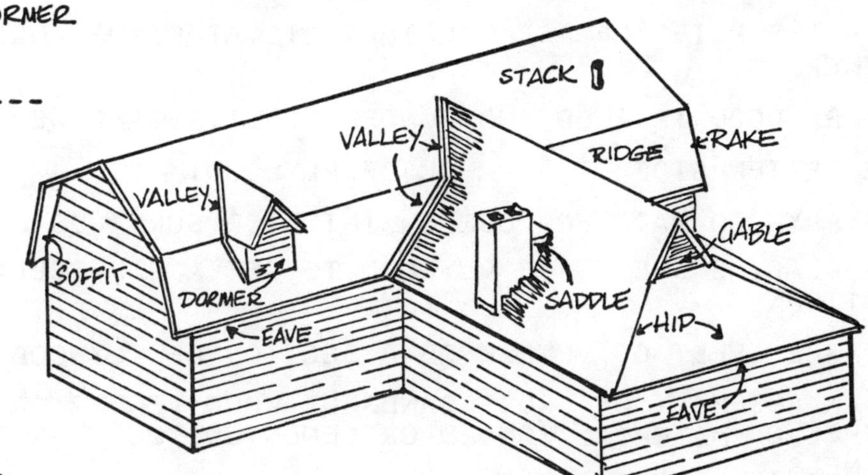

ROOFING TYPES-----

ALUMINUM, SHEET, SINGLES
ASBESTOS SHINGLES
ASPHALT SHINGLES
BUILT-UP
CANVAS DECK
CEMENT TILE
CLAY TILE
COPPER
FELT ASPHALT, TAR
FIBRE GLASS SHINGLES
GALVANIZED SHEET, SHINGLES
GRAVEL TOPPED
METAL TILE
ROLL ROOFING, SMOOTH, SURFACED
SLATE
STEEL COATED
STONE TOPPED
WOOD SHINGLES

"R" RESISTANCE VALUE-----

HEAT LOSS IS MEASURED IN "R" (RESISTANCE VALUE). USUALLY ALL INSULATING MATERIALS AND PRODUCTS HAVE AN "R" VALUE WHICH IS INDICATED ON THE PACKAGE OR CARTON OF THE NEW ITEMS.

THE HIGHER THE "R" VALUE THE GREATER THE INSULATION VALUE. THICKNESS OF AN INSULATOR IS NOT PROOF OF ITS INSULATION VALUE. YOU WILL REQUIRE A HIGHER "R" VALUE INSULATING MATERIAL THROUGHOUT YOUR HOME IF YOU LIVE IN THE STATES WITH COLDER WINTERS THAN YOU WOULD IF YOU LIVE IN THE SOUTHERN AREAS.

PLYWOOD TERMS-----

PLYWOOD IS USED IN ALMOST ALL TYPES OF MODERN CONSTRUCTION FOR BOTH INTERIOR AND EXTERIOR WORK. THESE ARE SOME OF THE TERMS COVERING THE VARIOUS TYPES OF THIS POPULAR PRODUCT.

- ADHESIVE - CEMENT-TYPE MATERIAL USED TO CEMENT THE PLIES TOGETHER; ALSO USED TO INSTALL MANY TYPES OF BUILDING PRODUCTS WITHOUT THE USE OF NAILS.
- BACK - THE POORER SIDE OF A PANEL OF PLYWOOD.
- CONSTRUCTION - ARRANGEMENT OF PLIES OR VENEERS IN THE PLYWOOD PANEL.
- CORE - THE CENTER OR INNERMOST PORTION OF THE PLYWOOD PANEL.
- CROSSBAND - THE PLIES OF A PANEL THAT LIE AT RIGHT ANGLES TO OTHERS.
- EXTERIOR - PLYWOOD TREATED AND BONDED TO WITHSTAND WEATHER.
- FACE - THE BETTER SIDE OF A PANEL OF PLYWOOD.
- INTERIOR - PLYWOOD MADE FOR USE ON INTERIOR SURFACES ONLY.
- MATCHING - PLACING SHEETS OF PLYWOOD TO GET A PARTICULAR GRAIN PATTERN.
- PLY - A SINGLE SHEET OF VENEER OF A SECTION OF PLYWOOD.
- PLYWOOD - AN ASSEMBLY OF CROSS BANDS OF WOOD VENEER OR OTHER TYPES OF MATERIAL BONDED OR CEMENTED TOGETHER TO MAKE PANELS.
- SCARFING - BEVEL CUTTING ENDS OF PLYWOOD PANELS FOR CEMENTING OR GLUING TOGETHER TO MAKE A GOOD JOINT.
- VENEER - THE THIN SHEETS OF WOOD BONDED TOGETHER TO MAKE PLYWOOD.
- WATERPROOF - THIS TERM IS USUALLY APPLIED TO EXTERIOR PLYWOOD.
- WATER RESISTANT - A PLYWOOD FOR LIMITED EXTERIOR USAGE.

SEE: *PLYWOOD, PLYWOOD MARKING, PLYWOOD SIDING, PLYWOOD WALL PANELING.*

ROOF TRUSSES-----

PREFABRICATED ROOF TRUSSES AND CANTILEVER BEAMS OF APPROVED TYPE, DESIGN AND STRENGTH ARE AVAILABLE IN MOST AREAS.

THEY ARE FACTORY BUILT IN MANY TYPES AND SIZES TO MEET MOST NORMAL CONDITIONS, CODES AND REGULATIONS.

TIME AND MONEY IS SAVED WHEN THEY ARE USED INSTEAD OF JOISTS, RAFTERS AND BEAMS CUT ON THE JOB AND BUILT IN PLACE.

THE ELEVATION BELOW SHOWS DETAILS OF A TYPICAL LOW SLOPE ROOF TRUSS FABRICATED USING METAL PLATE CONNECTORS AT THE VARIOUS JOINTS.

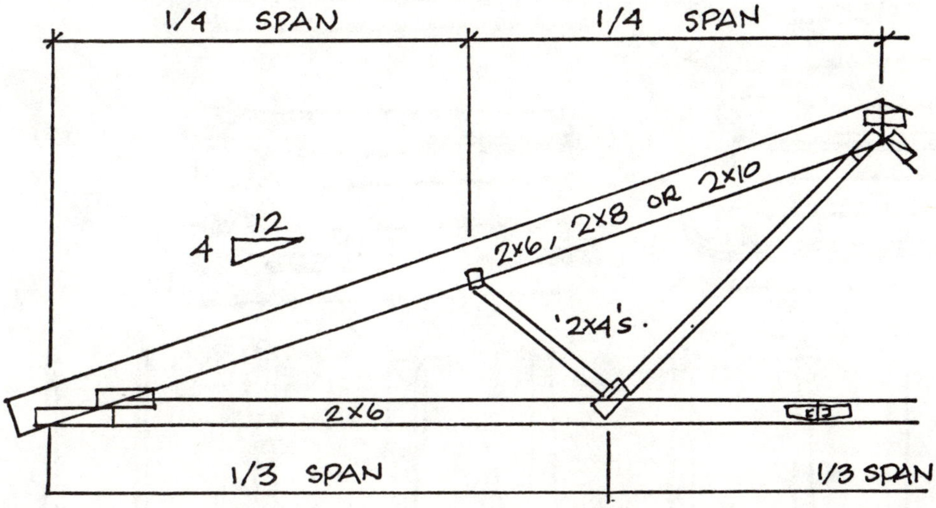

BELOW ARE SHOWN THREE TRUSS TYPES COMMONLY USED IN SMALL HOME CONSTRUCTION. THERE ARE MANY OTHER TRUSS TYPES SUITED TO ROOF STRUCTURE CONDITIONS OF ALL TYPES AND SIZES OF RESIDENTIAL BUILDINGS.

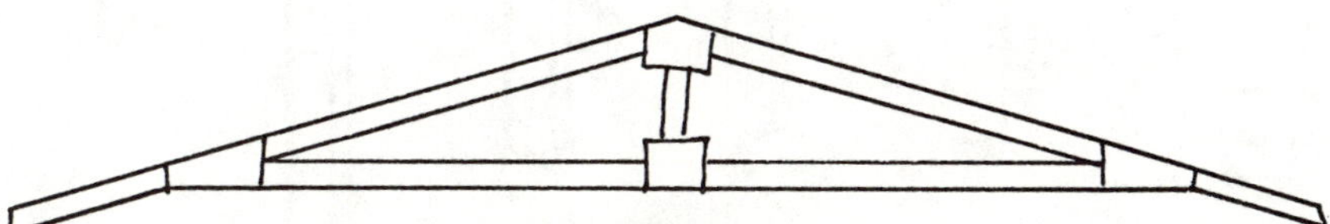

KINGPOST TRUSS

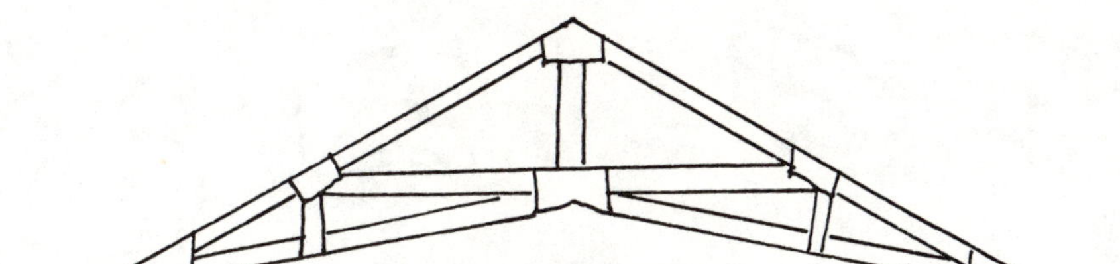

SCISSORS TRUSS

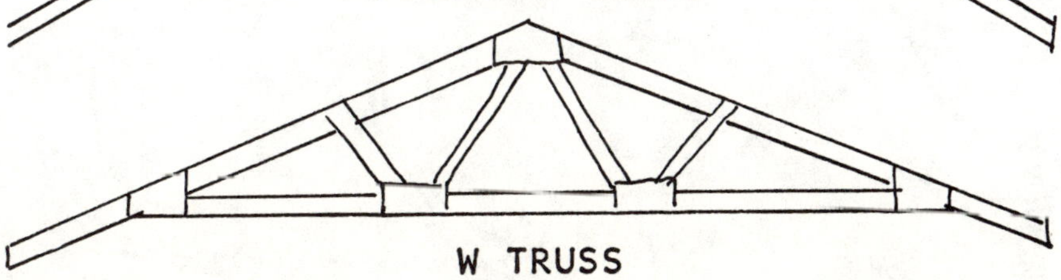

W TRUSS

STAIR PARTS-----

STAIRS ARE AN IMPORTANT AND INTEGRAL PART OF A GREAT MANY NEW HOMES AND OFTEN ARE INCLUDED WHEN REMODELING AND RESTORING HOMES. MANY ARE BUILT TO CONFORM TO MODERN AND PRESENT DAY LIVING AND DESIGN, BUT THEY STILL UTILIZE THE STANDARD PARTS SHOWN IN THESE SKETCHES, NAMED ON THE FOLLOWING PAGE.

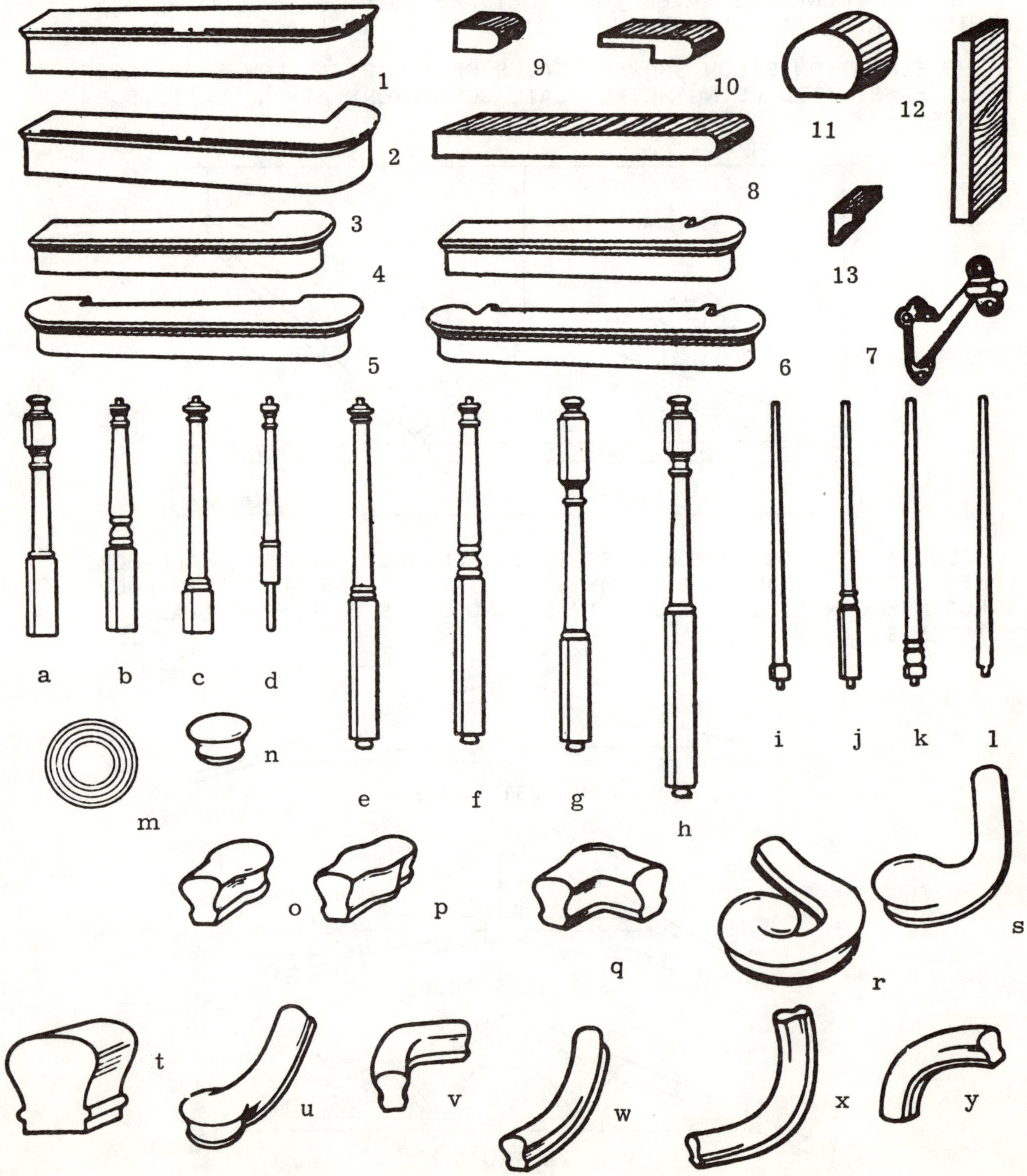

STAIR PARTS *(continued)*

THIS LIST OF STAIR PARTS COORDINATES WITH THE SKETCHES ON THE PREVIOUS PAGE.

1. 1/4 CIRCLE END STARTING STEP
2. 1/2 CIRCLE END STARTING STEP
3. BULLNOSE
4. SCROLL END STARTING STEP FOR VOLUTE
5. DOUBLE END NOSED STARTING STEP
6. DOUBLE SCROLL END STARTING STEP FOR VOLUTE
7. STAIR RAIL BRACKET
8. NOSED TREAD 1-1/16"X9-1/2" - 3'-3'6"-4'
9. RETURN NOSING 1-1/16"X1-1/4" X16"
10. LANDING TREAD 1-1/16"X3-1/4" - 3'-3'6"-4'
11. WALL RAIL 1-3/4"X1-5/8" - 8'-10'-12'-14'
12. STAIR RISER 3/4"X7-1/2" - 3'-3'6"-4'
13. STAIR COVE 5/8"X3/4"

A. STARTING NEWELL 3"X3"X48"
B-C. STARTING NEWELL 3-1/4" X3-1/4"X42"
D. VOLUTE NEWELL 2-1/4"X 2-1/4"X42"
E-F. LANDING NEWELL 3-1/4"X 3-1/4"X62"
G-H. LANDING NEWELL 3"X3"X62"
I-J-
K-L. TURNED BALUSTER 1-1/4"X 1-1/4" - 31"-34"-36"-39"-41"
M. WALL ROSETTE 3/4"X4-3/8"
N. NEWELL CAP 2-3/8"X4-1/4"
O. NEWELL CAP 1 RAIL LEVEL
P. NEWELL CAP 2 RAIL LEVEL
Q. NEWELL CAP RAIL 1/4 TURN
R. VOLUTE UP EASING
S. TURNOUT UP EASING
T. MAIN RAIL 2-1/4"X2-3/8"
U. START EASING - NEWELL CAP
V. LEVEL 1/4 TURN
W. REGULAR UP EASING
X. 90°
Y. EASING OVERHAND

TEMPLATES-----

TEMPLATES ARE USED TO SIMPLIFY SKETCHING OF TENTATIVE PLANS. THEY ARE AVAILABLE IN MOST OFFICE OR DRAFTING SUPPLY SHOPS.

THE MODERNIZER MODEL FOR KITCHENS AND BATHS SHOWN HERE IS 1/2" TO THE FOOT AND HAS CABINETS, COUNTERS, DOOR SWINGS, ELECTRIC SYMBOLS, AND PLUMBING FIXTURES.

THE FURNITURE MODEL IS 1/4" TO THE FOOT AND PROVIDES FOR FURNITURE SPACE AND ADAPTABILITY; COVERS OVER 50 PIECES OF FURNITURE WITH WIDTH, DEPTH AND HEIGHT OF EACH PIECE.

THE HOUSE PLAN MODEL IS A 1/4" SCALE SIMPLIFIED TEMPLATE FOR GENERAL USE WHEN YOU PLAN BUILDING, REMODELING, REPAIRING OR RESTORATION.

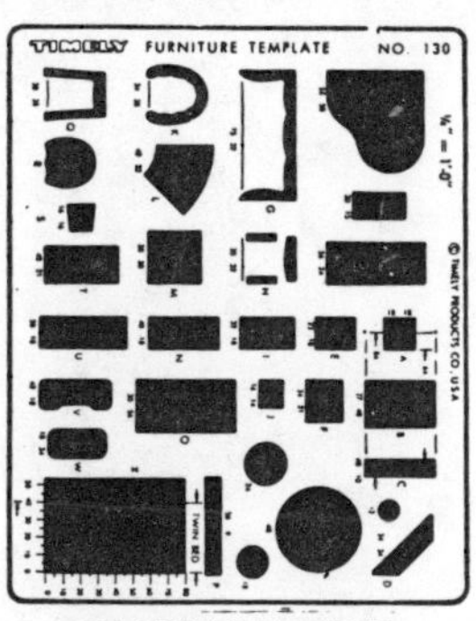
FURNITURE

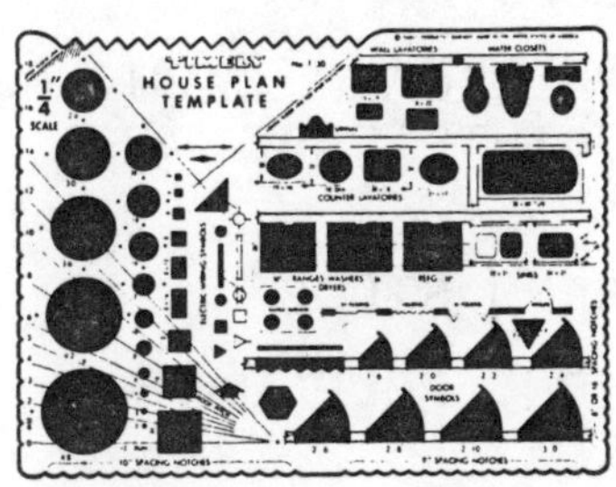
HOUSE PLAN

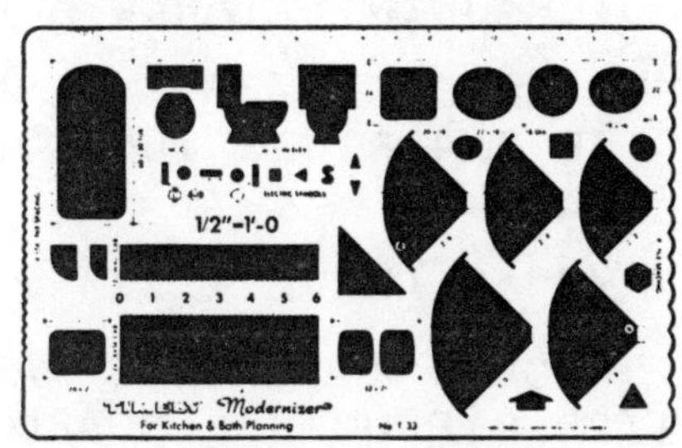
MODERNIZER

COURTESY TIMELY PRODUCTS CO.

SOUND INSULATION-----

TODAY MORE PEOPLE ARE CONSIDERING SOUND INSULATION IN HOMES THAN EVER BEFORE.

THE SOUNDS FROM TELEVISION SETS, RADIOS, STEREOS, MUSICAL INSTRUMENTS AND COMMON EVERY DAY LIVING NOISES NEED TO BE INSULATED FROM SEPARATE AREAS OF THE HOUSE.

LIKEWISE, EXTERIOR NOISE AND SOUNDS THAT MIGHT BE BOTHERSOME ALSO SHOULD BE ISOLATED FROM THE LIVING QUARTERS AND SLEEPING ROOMS. AS YOU CONSIDER THIS FACTOR, YOU WILL ENCOUNTER THE WORD "DECIBEL", WHICH IS THE TERM USED TO MEASURE SOUND ENERGY.

A DECIBEL IS THE SMALLEST CHANGE IN SOUND THAT THE AVERAGE EAR CAN DETECT.

SOUND INSULATION, LOSS-----

THE AVERAGE SOUND LOSS FOR VARIOUS TYPES OF WALL CONSTRUCTION IS LISTED BELOW.

CONSTRUCTION	AVERAGE SOUND LOSS (DECIBELS)
1/2" GYPSUM WALLBOARD EACH SIDE 2X4 STUDS 16" O.C., JOINTS TAPED AND FILLED	35
1/2" GYPSUM WALLBOARD EACH SIDE STAGGERED 2/3 STUDS 16" O.C., JOINTS TAPED AND FILLED	42
5/8" TYPE "X" GYPSUM WALLBOARD EACH SIDE 2X4 STUDS 16" O.C., JOINTS TAPED AND FILLED	37
SAME AS ABOVE - STAGGERED STUDS	44
TWO LAYERS 3/8" GYPSUM WALLBOARD CEMENTED TOGETHER EACH SIDE 2X4 STUDS 16" O.C., OUTER JOINTS TAPED AND FILLED	38
TWO LAYERS 1/2" GYPSUM WALLBOARD CEMENTED TOGETHER EACH SIDE 2X4 STUDS 16" O.C., OUTER JOINTS TAPED AND FILLED	43
TWO LAYERS 5/8" TYPE "X" WALLBOARD EACH SIDE 2X4 STUDS 16" O.C., JOINTS TAPED AND FILLED	40
SAME AS ABOVE - 2X3 STAGGERED STUDS	45
2" SOLID LAMINATED PARTITION, FOUR LAYERS 1/2" GYPSUM WALLBOARD	40
TWO LAYERS 3/8" GYPSUM WALLBOARD CEILING, 2X10 JOISTS 16" O.C., 1X4 FIR FINISH FLOOR AND 1X6 SUB-FLOOR OR 3/4" PLYWOOD	40

SKETCHING-----

IT'S EASY TO SKETCH FLOOR PLANS FOR KITCHENS, BATHS, ETC. USE THE SKETCH SHEETS FOUND IN THIS BOOK IN BOTH 1/4" AND 1/8" SCALE. YOU MAY ALSO WANT TO CONSIDER USING TEMPLATES. THE SKETCH AT RIGHT WILL GIVE YOU SOME IDEAS ON HOW TO START.

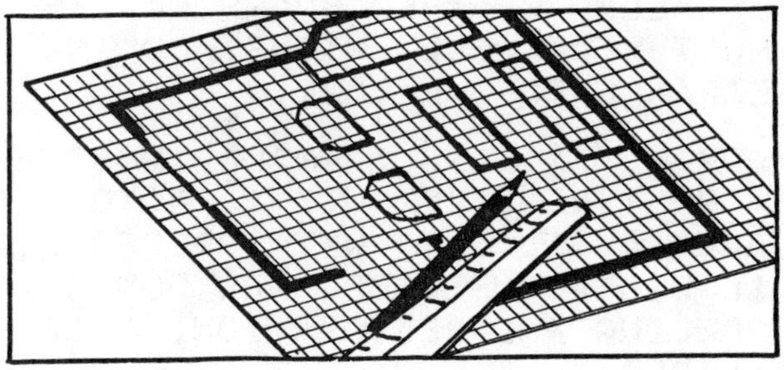

SEE: *TEMPLATES, FURNITURE SKETCHES, FURNITURE TYPES, SIZES.*

SECOND MORTGAGES-----

THE SECOND MORTGAGE ON REAL ESTATE IS AN INSTRUMENT TO PROVIDE ADDITIONAL FINANCING. IT MIGHT COVER SOME OR ALL OF THE COST FOR MODERNIZATION, PART OF A DOWN PAYMENT, ETC. THE FIRST MORTGAGE ALWAYS TAKES PRECEDENCE IN THE CASE OF FAILURE TO PAY AND THE SECOND MORTGAGE HOLDER CANNOT REDEEM HIS CLAIM UNTIL THE FIRST MORTGAGE HOLDER IS SATISFIED. IN SOME CASES, A FIRST AND SECOND MORTGAGE IS PART OF AN ORIGINAL DEAL.

ALL MORTGAGES, EASEMENTS, RIGHTS OF WAY, PROPERTY LINES, SUITS, ETC., ARE FILED AND RECORDED IN THE PUBLIC OFFICES OF THE COUNTY OR OTHER PUBLIC JURISDICTION AND ARE OPEN TO THE PUBLIC OR AUTHORIZED PERSONS FOR CHECKING.

TREATED LUMBER-----

MANY KINDS OF LUMBER MAY BE TREATED WITH PRESERVATIVES TO PROTECT IT AGAINST DECAY, DRY ROT, DAMPNESS, TERMITES AND OTHER DESTRUCTIVE FORCES. PRESERVATIVES ARE AVAILABLE FOR LUMBER OR WOOD USED ABOVE GROUND AND FOR LUMBER PRODUCTS WHICH COME IN CONTACT WITH THE GROUND.

THE THREE CLASSES AND TYPES OF PRESERVATIVES ARE WATER BORNE SALT, CREOSOTE AND COAL TAR SOLUTIONS, AND CHLOROPHENOL. USUALLY WOOD PRODUCTS TREATED THIS WAY ARE READY TO USE FOLLOWING PRESSURE TREATMENTS IN SPECIAL PLANTS. IN SOME CASES CERTAIN TREATMENTS MAY BE APPLIED BY BRUSHING OR DIPPING.

IF YOUR PLANS INCLUDE WORK WHERE WOOD PRODUCTS WILL BE SUBJECTED TO THESE CONDITIONS, CHECK INTO THE USE OF TREATED LUMBER OR USE ONE OF THE LUMBER PRODUCTS WHICH DO NOT REQUIRE TREATMENT.

SIDING, METAL, VINYL-----

WHETHER YOU ARE CONSIDERING BUILDING, REMODELING OR RESTORING A HOME, THE PROBLEM OF SIDING WILL BE IMPORTANT. SIDING (AT ONE TIME CALLED WEATHER BOARDING) IS THE OUTSIDE COATING OF A HOUSE. IT MUST WEAR WELL, WITHSTAND THE ELEMENTS, AND GIVE A GOOD APPEARANCE TO THE HOME.

ALUMINUM SIDING IS MADE IN SEVERAL TYPES, SIZES AND TEXTURES TO MEET MOST REQUIREMENTS. FACTORY COATING IN SEVERAL COLORS IS USUALLY AVAILABLE, WITH PROPER FITTINGS FOR INSTALLATION TO MATCH. IT IS WIDELY USED FOR NEW HOMES AS WELL AS COVERING WORN WALLS IN REMODELING AND RESTORATION. COATED STEEL SIDING MAY ALSO BE AVAILABLE.

VINYL AND VINYL-COATED WOOD OR COMPOSITION MAY BE FOUND IN SEVERAL COLORS AND TYPES FOR GENERAL USE. THIS TYPE MAY BE A SATISFACTORY SOLUTION WHERE POLLUTION AND CORROSIVE CONDITIONS MIGHT AFFECT PAINTS OR OTHER SURFACES.

HORIZONTAL TYPES

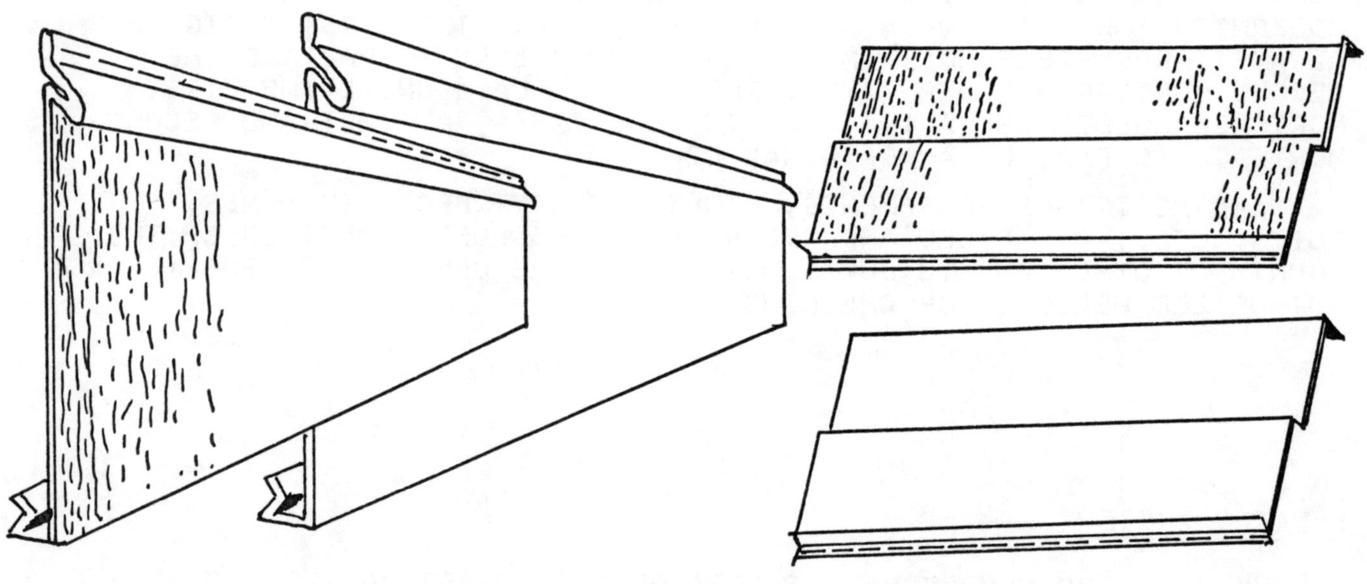

VERTICAL TYPES

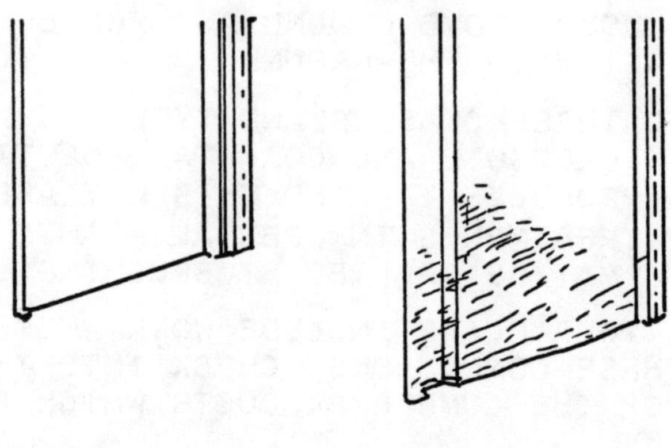

SIDING, METAL, VINYL (continued)

SOME OF THE PATTERNS AND DESIGNS IN ALL TYPES RESEMBLE LAP SIDING. SMOOTH SURFACES, AS WELL AS TEXTURES RESEMBLING WOOD GRAIN AND ROUGH SAWED WOOD, MAY BE FOUND IN SUPPLIER'S STOCKS.

VERTICAL PATTERNS OF VARIOUS DESIGNS ARE MADE BY SOME PRODUCERS. IN MOST CASES, LONG LIFE, FREEDOM FROM DECAY, CLEANING AND INFREQUENT PAINTING ARE THE ADVANTAGES OF THESE MATERIALS. WHEN REQUIRED, THESE PRODUCTS ARE DESIGNED TO BE USED IN CONJUNCTION WITH INSULATION MATERIALS. OTHER TYPES OF SIDING TO CONSIDER ARE PLYWOOD AND FACTORY FINISHED HARDBOARD OR COMPOSITION TYPES.

THE SKETCHES ON THE PREVIOUS PAGE SHOW SOME TYPES AND FINISHES AVAILABLE IN MOST AREAS.

SIDING, WOOD-----

SOLID WOOD OR NATURAL WOOD SIDINGS ARE GRADUALLY BECOMING A PART OF THE PAST. AT ONE TIME, IT WAS COMMON FOR SAWMILLS ALL AROUND THE COUNTRY TO TURN OUT SIDING BOARDS IN VARIOUS REGIONAL WOOD SPECIES AND IN A WIDE RANGE OF PATTERNS OR STYLES. THE MOST COMMON WERE DROP, SHIPLAP AND BEVEL SIDING SHAPES. THAT PRACTICE HAS CHANGED AS WOOD SIDING BEGAN LOSING THE COMPETITIVE BATTLE ON NEW HOME CONSTRUCTION TO OTHER SIDING MATERIALS COSTING LESS AND EASIER TO APPLY.

IN MANY CASES, WOOD SIDINGS CUT IN FLAT GRAIN STYLE GAVE WAY TO EDGE GRAIN SIDINGS WHICH HAVE PROVED SUPERIOR IN THEIR PAINT HOLDING PROPERTIES. TODAY, SIDING BOARDS (FOR THE MOST PART) ARE LIMITED IN THEIR PATTERNS AND ARE OFFERED PRIMARILY IN PINE, WESTERN RED CEDAR AND REDWOOD SPECIES.

ANOTHER FACTOR IN THE SIDING TREND PICTURE HAS BEEN THE RISE IN PREFERENCE FOR PREFINISHED SIDINGS AMONG NEW HOME BUILDERS AND HOME REMODELERS. THIS PREFERENCE HAS BROADENED AS NEW AND BETTER FACTORY FINISHES HAVE BEEN DEVELOPED. THIS GROWTH IN SOPHISTICATED PREFINISHES, BEGUN BY METAL AND HARDBOARD SIDING PRODUCERS, HAS SPREAD TO PLYWOOD AND FIBERBOARD MATERIALS. ALSO, MANY OF THE NEWER PREFINISHES INCLUDE SURFACE TEXTURING THAT SIMULATES WOOD GRAIN, SAWING MARKS OR OTHER DETAILS.

STILL, THE LONG TRADITIONAL NATURAL WOOD MATERIALS REMAIN FAVORITES OF MANY HOME OWNERS, REMODELERS AND BUILDERS. THIS IS ESPECIALLY TRUE IN CERTAIN AREAS SUCH AS THE PACIFIC NORTHWEST, CALIFORNIA, THE MID-SOUTH, AND PARTS OF THE NORTHEAST WHERE LUMBERING HAS FLOURISHED A LONG TIME.

SELECTING A HOMESITE-----

THERE ARE MANY THINGS TO CONSIDER WHEN YOU SELECT THE LOCATION AND SITE FOR YOUR NEW HOME. IF YOU HAVE A FAMILY OF SCHOOL-AGE CHILDREN, THE PROXIMITY OF THE SCHOOLS WILL BE VERY IMPORTANT. YOU SHOULD THEN CHECK, IN ORDER, THE DISTANCE OR TRAVELING TIME TO YOUR PLACE OF EMPLOYMENT, AVAILABILITY OF MARKET AREAS, PUBLIC TRANSPORTATION IF REQUIRED, THROUGH STREETS, HIGHWAYS AND EXPRESS-WAYS.

AFTER YOU CHECK THESE FACTORS YOU SHOULD CONSIDER THE ZONING REGULATIONS, BUILDING CODES AND RESTRICTIONS, EASEMENTS, AND UTILITY SERVICES—GAS, ELECTRICITY, PHONE, TV CABLE, WATER, GAR-BAGE, ETC. IF YOU HAVE HEALTH PROBLEMS, THE POLLUTION IN THE AREA MAY REQUIRE CHECKING ALSO.
POLICE AND FIRE PROTECTION AND INSURANCE RATES FOR THE TYPE OF HOME YOU EXPECT TO BUILD. ALL OF THESE ITEMS WILL ENTER INTO THE PRESENT AND FUTURE VALUE AND LIVABILITY OF YOUR HOME.

BEFORE YOU AGREE TO BUY A SITE OR LOT, A FEW OF THE SITE'S PHYSICAL FEATURES SHOULD BE CHECKED FOR "OFF SITE IMPROVEMENTS"— STREETS, WALKS, STORM AND SANITARY SEWERS. IF THEY ARE NOT IN-CLUDED IN THE PURCHASE PRICE OR PRESENT TAX ASSESSMENTS, THEY MAY BE ASSESSED LATER ON. IT PAYS TO CHECK ALL FUTURE TAX ASSESSMENTS WHEN PURCHASING ANY BUILDING SITE.

SOLAR HEATING-----

DUE TO THE IMPORTANCE OF ENERGY TODAY, SOLAR HEATING IS NOW BEING CONSIDERED VERY SERIOUSLY IN MANY AREAS OF THE COUNTRY FOR HOME HEATING AND HOT WATER SUPPLY. IN THE AREAS WHERE SUNSHINE PREDOMINATES MOST OF THE DAYLIGHT HOURS IT APPEARS TO HAVE A PRACTICAL APPLICATION FOR THIS PURPOSE.

IF YOU LIVE IN AN AREA OF THE COUNTRY WHERE THIS METHOD SEEMS FEASIBLE, YOU CAN CONSIDER THIS METHOD OF SAVING ENERGY AND MONEY WHEN YOU BUILD, REMODEL OR RESTORE A HOME. YOU SHOULD GET ALL THE DATA AVAILABLE FROM YOUR STATE ENERGY AGENCIES AND FROM THE NATIONAL SOLAR HEATING & COOLING INFORMATION CENTER, P.O. BOX 1607, ROCKVILLE, MD 20850.

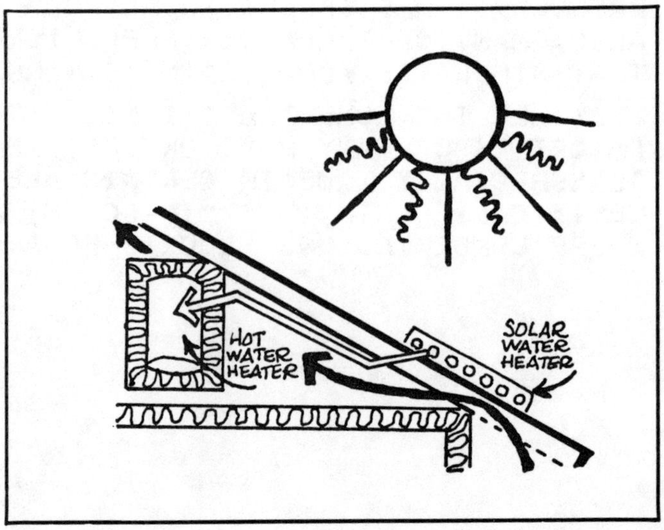

SPECIFICATIONS-----

SPECIFICATIONS, COMMONLY KNOWN IN THE BUILDING INDUSTRY AS "SPECS", SHOW OR GIVE EXACT INFORMATION ON PRODUCTS, MATERIALS, WORKMANSHIP AND MANY DETAILS THAT ARE NOT SHOWN ON THE PLANS FOR A HOUSE. HOUSE PLANS SHOW SIZES, DIMENSIONS, CONSTRUCTION DETAILS AND INFORMATION ON MATERIALS INSOFAR AS POSSIBLE. HOWEVER, A GREAT MANY IMPORTANT DETAILS AND INFORMATION CANNOT BE INCLUDED IN ALL PLANS.

FORMAL OR REGULATION TYPE "SPECS" ARE USUALLY USED ON NEW HOUSE JOBS, AND MAY BE INCLUDED AS PART OF THE CONTRACT OR AGREEMENT. ON MODERNIZATION OR RESTORATION WORK, THE SPECIFICATIONS ARE OF THE CASUAL TYPE AND MAY BE IN THE FORM OF A LETTER, PART OF THE CONTRACT, OR PRICE FOR THE WORK.

WHEN YOU BUY A SET OF PLANS A GOOD PORTION OF THE SPECIFICATION DETAILS ARE SHOWN ON THE PLANS THEMSELVES. HOWEVER, A SET OF "SPECS" WITH BLANK SPACES FOR INSERTING LOCAL CONDITIONS AND/OR YOUR CHOICE OF MATERIALS, PRODUCTS OR SERVICES CAN BE PROVIDED FOR YOUR USE.

TAXES, REAL ESTATE-----

TAXES ARE ASSESSED AND COLLECTED PERIODICALLY BY THE CITY, COUNTY, STATE OR OFFICE ESTABLISHED AND AUTHORIZED TO HANDLE THIS MATTER. THEY ARE PAYABLE PROMPTLY ON THE DATE SET WITH A PENALTY INDICATED FOR LATE PAYMENT. PROLONGED DELINQUENCY RESULTS IN FORECLOSURE AND SALE OF THE PROPERTY FOLLOWING DUE NOTICE TO THE OWNERS. IN MANY CASES THE TAX PAYMENTS ARE HANDLED DIRECTLY BY THE LENDING INSTITUTION WHICH HAS MADE THE LOAN—WITH THE TAXES ADDED TO THE MONTHLY PAYMENTS MADE ON THE LOAN. IF YOU ARE BUYING PROPERTY, THE SELLER OR AGENT USUALLY PROVIDES COPIES OF PAID TAX BILLS TO SHOW THE PRESENT TAX CHARGES. AS A RULE, PRESENT PRACTICES GENERALLY PROVIDE FOR A SIX MONTH PRORATING OF TAXES. TAX AND INSURANCE TERMS AND CONDITIONS SHOULD BE COVERED IN THE SALES AGREEMENT. IN THE CASE OF A NEW HOME, THE TAXES ONLY COVER THE ACTUAL REAL ESTATE AND THE EVALUATION COVERING THE HOME IS NOT MADE UNTIL THE STRUCTURE IS COMPLETED. IN EITHER CASE, THE BUYER OR TITLE CHECKER SHOULD ANTICIPATE ANY FUTURE TAX ASSESSMENTS THAT MAY BE NECESSITATED BY IMPROVEMENTS OR LEGISLATION.

Notes: ____________________

TAX AND INSURANCE PRORATING-----

WHEN A PROPERTY IS BOUGHT OR SOLD THE TAXES AND/OR THE INSURANCE ARE USUALLY PRORATED ON A SIX MONTHS BASIS. (THIS PRACTICE MAY VARY IN SOME AREAS—OR THE SALES AGREEMENT OR CONTRACT MAY PROVIDE OTHER ARRANGEMENTS.) WHERE THIS METHOD PREVAILS, THE SELLER USUALLY PAYS THE PRORATED AMOUNT FOR EITHER OR BOTH ITEMS. YOU CAN EASILY DETERMINE THE AMOUNT IN QUESTION BY THE USE OF THIS TABLE AND THE EXAMPLE BELOW.

1 MONTH----.1666¢	4 MONTHS----.6666¢
2 MONTHS---.3333¢	5 MONTHS----.8333¢
3 MONTHS---.50¢	6 MONTHS--$1.00

1 DAY---.0055¢	11 DAYS---.0611¢	21 DAYS---.1166¢
2 DAYS--.0111¢	12 " ---.0666¢	22 " ---.1221¢
3 " ---.0166¢	13 " ---.0722¢	23 " ---.1277¢
4 " ---.0222¢	14 " ---.0777¢	24 " ---.1333¢
5 " ---.0277¢	15 " ---.0833¢	25 " ---.1389¢
6 " ---.0333¢	16 " ---.0888¢	26 " ---.1444¢
7 " ---.0388¢	17 " ---.0944¢	27 " ---.15¢
8 " ---.0444¢	18 " ---.0999¢	28 " ---.1555¢
9 " ---.05¢	19 " ---.1055¢	29 " ---.1611¢
10 " ---.0555¢	20 " ---.1111¢	30 " ---.1666¢

FOR EXAMPLE, ASSUME TAXES ARE PRORATED ON A 6 MONTHS BASIS STARTING ON OCTOBER 20TH. THUS YOU WOULD HAVE JULY, AUGUST AND SEPTEMBER—OR 3 MONTHS OF THE SECOND HALF OF THE YEAR—PLUS 20 DAYS. REFERRING TO THE TABLE YOU WILL FIND 3 MONTHS EQUALS 50¢ AND 20 DAYS EQUALS .1111¢—A TOTAL OF .6111¢ FOR EACH $1 OF TAX AND/OR INSURANCE TO BE PAID BY THE SELLER. THE AMOUNT SHOWN IS PER $1 OF TAXES AND INSURANCE.

TITLES-----

INFORMATION ON TITLES, LIENS, MORTGAGES, LAND CONTRACTS, ETC. MAY BE OBTAINED FROM THE COUNTY WHERE THE PROPERTY IS LOCATED. IN SOME AREAS THIS MAY BE HANDLED BY THE CITY OR STATE DEPARTMENTS. IF YOU ARE BUYING PROPERTY, THE SELLER FURNISHES A CLEAR TITLE—WHICH IS ALSO REQUIRED BY THE ORGANIZATION FURNISHING THE FINANCING. IF A CASH DEAL IS PROPOSED, THE BUYER SHOULD OBTAIN THE CLEAR TITLE FROM THE SELLER. A TITLE CONCERN WILL HANDLE THIS AS A RULE, AND ALSO WILL PROVIDE A GUARANTEED OR INSURED TITLE. LEGAL ADVICE SHOULD ALWAYS BE CONSIDERED IN SITUATIONS WHERE A LENDING INSTITUTION IS NOT AVAILABLE TO ASSIST THE BUYER IN CHECKING THE VARIOUS PHASES OF THE TRANSACTION.

<u>TERMITES</u>-----

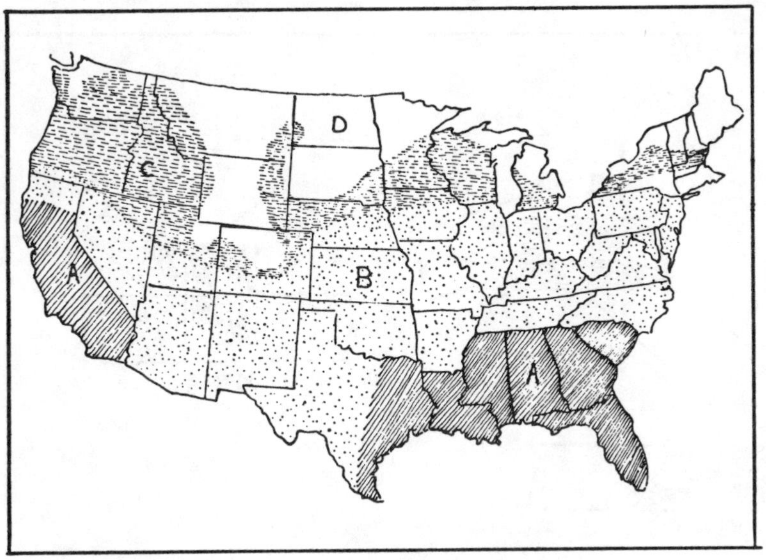

TERMITE SUSCEPTIBILITY BY GEOGRAPHIC AREA
A. Region I (including Hawaii): termite protection required.
B. Region II: termite protection generally required, although specific areas sometimes exempted.
C. Region III: termite protection usually not required, except specific local areas.
D. Region IV (including Alaska): termite protection not required.
SOURCE: U.S. Department of Housing and Urban Development (HUD) Minimum Property Standards. *Washington, D.C.: Government Printing Office.*

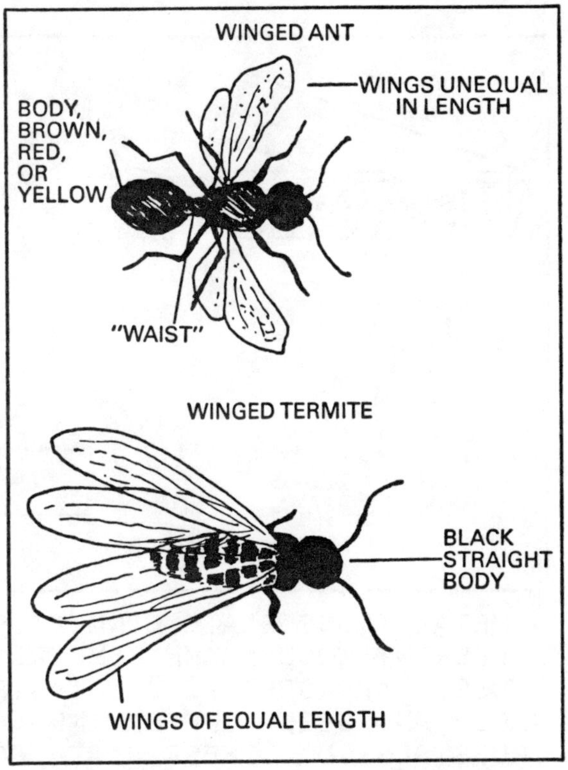

Winged ants are commonly mistaken for winged termites. Differences are noted in sketch.

AS SHOWN BY THE ABOVE MAP, TERMITES MAY BE FOUND IN A GREAT MANY PARTS OF THE COUNTRY. IF YOU CONTEMPLATE NEW CONSTRUCTION, TREATED LUMBER MAY BE USED AT THE MOST CRITICAL POINTS TO AVOID THE THREAT OF THESE INSECTS.

IF THEY ARE FOUND IN PRESENT BUILDINGS, PROTECTION OR ELIMINATION MAY BE AVAILABLE THROUGH THE SERVICES OF PEST CONTROL CONCERNS WHICH OPERATE AGENCIES THROUGHOUT THE COUNTRY.

VACATION OR SECOND HOMES IN REMOTE AREAS THAT ARE OCCUPIED ONLY PART OF EACH YEAR ARE MOST SUSCEPTIBLE TO TERMITE DAMAGE AND REQUIRE SPECIAL ATTENTION.

MANY TIMES WINGED ANTS OR OTHER INSECTS ARE MISTAKEN FOR TERMITES. THE ABOVE SKETCH WILL ENABLE YOU TO IDENTIFY THE TERMITE.

THEY ARE PREVALENT IN AREAS WHERE HIGH HUMIDITY, WARMER TEMPERATURES AND SOFT, DAMP SOIL CONDITIONS ARE FOUND.

TERMITES EAT AND LIVE ON WOOD. THEY ARE FOUND MOSTLY IN WOOD THAT IS DECAYED OR STARTING TO DECAY.

FROM THEIR NESTS (WHICH MAY EVEN BE BELOW THE FROST LINE) THEY BUILD MUD TUNNELS TO REACH THE WOOD IN A STRUCTURE.

SEE: *BUYING A HOME, TREATED LUMBER.*

TRUSS PLATE CONNECTORS-----

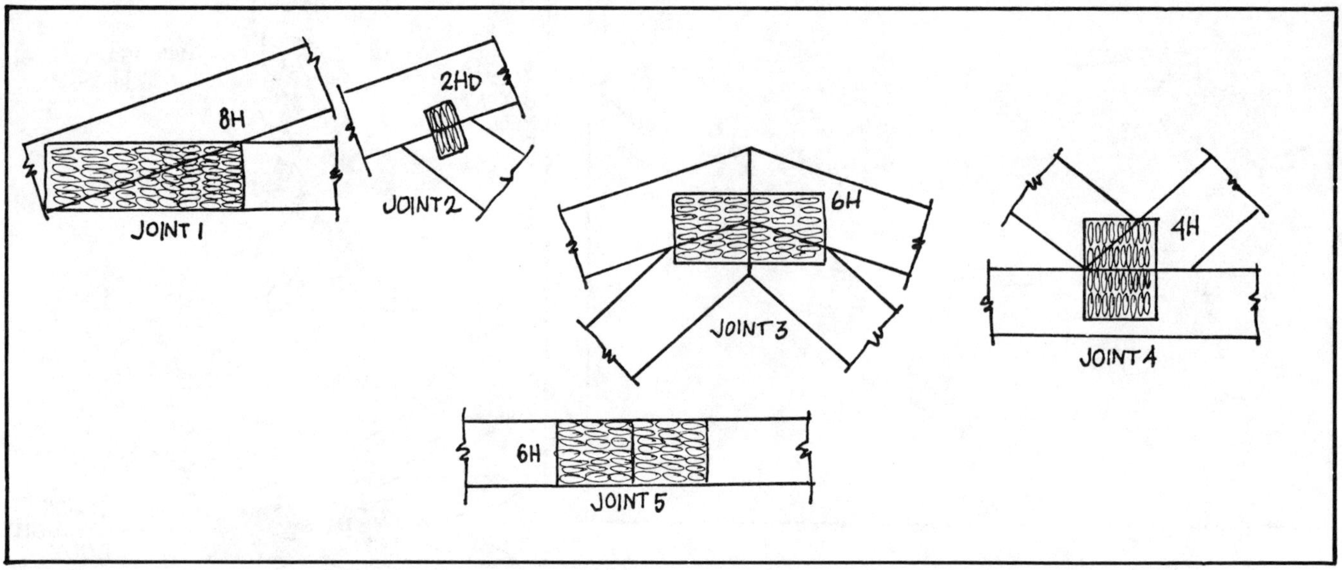

METAL TRUSS PLATE CONNECTORS ARE USED TO SECURELY JOIN THE DIFFERENT WOOD FRAME MEMBERS THAT COMPRISE A ROOF OR FLOOR TRUSS. APPROPRIATE JOINT STRENGTH IS ASSURED BY PROPER PLATE GAUGE AND SIZE - DETERMINED BY AN ENGINEERING ANALYSIS OF THE TRUSS DESIGN. THESE PLATE CONNECTED TRUSSES ARE FABRICATED IN MORE THAN 2,000 COMPONENTS PLANTS SCATTERED THROUGHOUT THE COUNTRY.

SEE: *ROOF TRUSSES*. ALSO, SUCCESSFUL PUTTING IT ALL TOGETHER (STRUCTURES PUBLISHING COMPANY)

Notes:

WOOD CONNECTORS AND CLIPS IN HOUSE CONSTRUCTION, USES-----

METAL WOOD CONNECTORS AND CLIPS ARE NOW USED EXTENSIVELY IN HOUSE AND LIGHT CONSTRUCTION. THEY SAVE TIME, MATERIAL AND MONEY AND PROVIDE STRONG CONSTRUCTION AND HIGH QUALITY WORKMANSHIP.

THE FRAMING AND CONSTRUCTION SKETCH AT THE RIGHT SHOWS THE LOCATION AND USES OF THE VARIOUS KINDS OF CONNECTORS AND CLIPS, AS ADAPTED TO MODERN CONSTRUCTION.

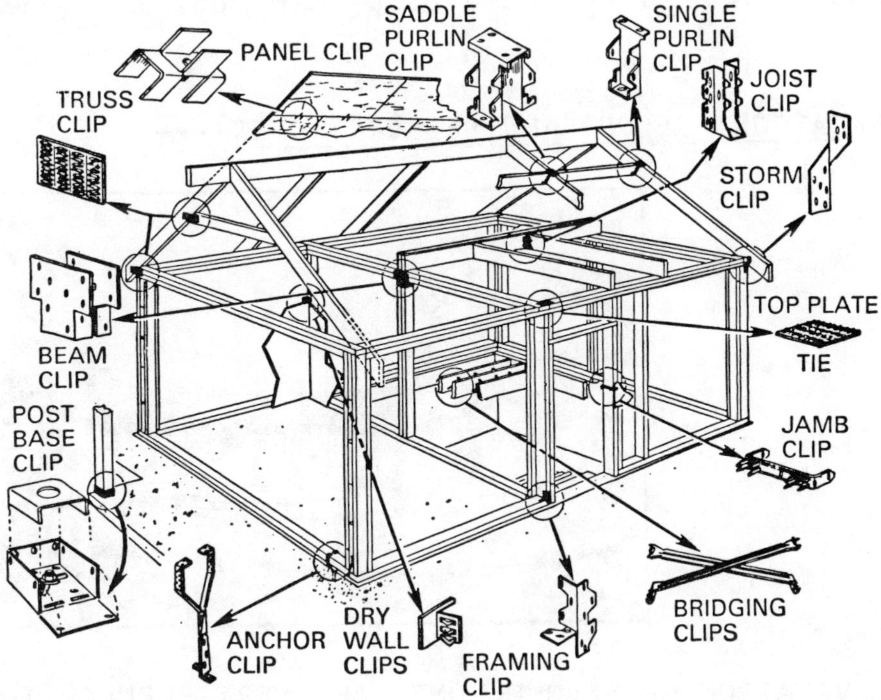

COURTESY: PANEL CLIP COMPANY,
FARMINGTON, MICHIGAN

SEE: *CLIPS, CONNECTORS*.
ALSO- SUCCESSFUL PUTTING IT ALL TOGETHER (STRUCTURES PUBLISHING)

SURVEYS-----

DEEDS FOR REAL ESTATE ONLY DESCRIBE THE "REAL PROPERTY" OR THE LAND. SURVEYS ON ALL PROPERTY GIVE THE DESCRIPTION, SIZE, RESTRICTIONS, ETC., OF THE PLOT. SURVEYS AND NEW DESCRIPTIONS ARE NECESSARY WHEN PLOTS ARE DIVIDED INTO NEW SIZES. AS A RULE, THIS MUST BE DONE BY A REGISTERED SURVEYOR OR ENGINEER.

UNDERWRITERS' LABORATORIES-----

THE UNDERWRITERS' ORGANIZATION IS SPONSORED BY THE INSURANCE INDUSTRY TO CONTROL PRODUCTS, MATERIALS AND WORKMANSHIP THAT AFFECT SAFETY, FIRE CONTROL, PROPERTY DAMAGE AND HUMAN SAFETY. ROOFING, ELECTRICAL WIRING, APPLIANCES AND INSTALLATION, AS WELL AS OTHER BUILDING COMPONENTS, ARE UNDER THE JURISDICTION OF THIS ORGANIZATION. THEIR LABELS AND CLASSIFICATION RATINGS MAY BE FOUND ON MANY PRODUCTS USED IN HOMES.

VACATION, SECOND AND LEISURE HOMES-----

VACATION OR SECOND HOMES ARE VERY POPULAR TODAY AND THE DEMAND IS GROWING FAST. HOMES OF THIS TYPE MAY BE MOUNTAIN CHALETS, BEACH HOUSES, SEASHORE HOMES, LAKE COTTAGES OR MANY OTHER TYPES AND DESIGNS. THEY MAY BE SMALL AND SIMPLE WITH MINIMAL LIVING ACCOMMODATIONS, OR LARGE COMPLETE HOMES WITH ALL THE AMENITIES OF MODERN LIVING.

GENERALLY SPEAKING, YOU WILL FIND THE SAME CONDITIONS FOR HOME BUILDING APPLY TO VACATION HOMES. MORTGAGE CONDITIONS AND INSURANCE MAY VARY DUE TO REMOTE LOCATIONS. UTILITIES, WATER AND SANITARY CONVENIENCES WILL ALSO DEPEND ON THE LOCATION. MATERIALS AND OTHER ITEMS MAY BE LIMITED AND YOUR PLANS MAY BE DIFFICULT AND EXPENSIVE TO FULFILL.

Notes:

WAIVER OF LIEN-----

THE WAIVER IS NECESSARY FOR YOU AND THE LENDING AGENCY WHEN MONEY IS BORROWED ON A BUILDING PROJECT. IT WILL PROTECT YOU AND THE LENDER AGAINST MECHANICS' LIENS THAT MIGHT BE PLACED AGAINST LABOR OR MATERIALS FURNISHED FOR THE JOB. THIS FORM IS A LEGAL TYPE THAT IS GENERALLY USED. IT SHOULD BE EXECUTED PROPERLY BEFORE PAYMENTS ARE MADE ON ANY PORTION OF A JOB.

State of ______________ } ss.

County of ______________ }   ______________ 19____

TO ALL WHOM IT MAY CONCERN:

Whereas ______ the undersigned ______________

ha____ been employed by ______________

to furnish ______________

____

for the Building known as ______________

City of ______________

____

Lot No. ____________ Section ____________ Township ____________ Range ________

County of ______________ State of ______________

NOW, THEREFORE, KNOW YE, That ______________ the undersigned for and in consideration of the sum of ______________ Dollars and other good and valuable considerations, the receipt whereof is hereby acknowledged, do hereby waive and release any and all lien, or claim or right to lien on said above described building and premises under the Statutes of the State of ____________ relating to Mechanics' Liens, on account of labor or materials, or both, furnished or which may be furnished, by the undersigned to or on account of the said ______________

______________ for said building or premises.

Given under____ hand____ and seal____ this ______________ Day of ____________ A. D., 19____

Witness: ______________, ______________ (SEAL)

______________ ______________ (SEAL)

SEE: *MECHANICS LIENS.*

Notes: ______________

____

____

____

____

____

____

WALLPAPER-----

WALLPAPER IS NOW USED IN MANY ROOMS OF THE MODERN HOME AND LENDS MUCH TO THE DECOR, LIVABILITY AND COMFORT OF THE HOME.

THE ACCOMPANYING TABLE WILL ENABLE YOU TO QUICKLY CALCULATE THE WALLPAPER REQUIREMENTS FOR VARIOUS SIZED ROOMS AND CEILING HEIGHTS.

Room Dimensions	*Height of Ceiling*				*No. Rolls for Ceiling*
	7 Ft.	*8 Ft.*	*9 Ft.*	*10 Ft.*	
6 X 10	8	9	10	11	2
6 X 12	8	10	11	12	3
8 X 10	9	10	11	12	3
8 X 12	9	11	12	13	3
9 X 10	9	11	12	13	3
9 X 12	10	12	13	16	4
9 X 14	11	12	14	16	4
9 X 16	12	13	15	17	5
10 X 10	9	11	12	13	4
10 X 12	10	12	13	15	4
10 X 14	11	13	15	16	5
11 X 11	10	12	13	15	4
11 X 12	11	12	14	16	4
11 X 16	13	14	16	18	6
14 X 14	13	15	17	19	7
16 X 18	16	18	20	23	9
16 X 20	17	19	22	24	10

THE NUMBER OF 24" WIDTH, READY TRIMMED ROLLS IS INDICATED IN THE COLUMNS.

DEDUCT A HALF ROLL FOR EACH AVERAGE SIZED DOOR, WINDOW OR EQUIVALENT SIZED OPENING IN THE ROOM.

SOME TYPES OF WALLPAPER COME IN DOUBLE ROLLS WHICH SAVES WASTE IN PATTERN MATCH-UP.

SEE: <u>SUCCESSFUL WALL COVERINGS AND DECORATION</u> (STRUCTURES PUBLISHING COMPANY)

Notes:

WATER CONDITIONING-----

WITH THE INCREASING IMPORTANCE OF ECOLOGY, OUR WATER PROBLEMS BECOME GREATER EACH DAY. WATER FOR YOUR HOME MUST BE CONSIDERED FROM ALL ANGLES. THESE ARE SOME OF THE COMMON TERMS DEALING WITH WATER.

- CALCIUM - ONE OF THE PRINCIPAL ELEMENTS FOUND IN THE EARTH'S CRUST. COMPOUNDS OF THIS ELEMENT, WHEN DISSOLVED IN WATER, MAKE IT HARD. CALCIUM IN WATER IS A FACTOR CONTRIBUTING TO THE FORMATION OF SCALE AND INSOLUBLE SOAP CURDS—TWO INDICATIONS OF HARD WATER. IT IS SOMETIMES REFERRED TO AS "LIME."
- CALCIUM CARBONATE EQUIVALENT - ALL FORMS OF HARDNESS AND OTHER SALTS ARE COMMONLY EXPRESSED IN TERMS OF CALCIUM CARBONATE EQUIVALENTS.
- CORROSION - THE DISINTEGRATION OF METALS BY ELECTROCHEMICAL MEANS. CORROSION OF IRON OR STEEL IS COMMONLY CALLED RUSTING.
- DETERGENT - TECHNICALLY, ANY TYPE OF CLEANSING SUBSTANCE MAY BE CALLED A DETERGENT. THE TERM IS MOST FREQUENTLY USED IN REFERENCE TO CLEANSERS WHICH FOAM AND ACT LIKE SOAP BUT ARE NOT MADE FROM FATTY ACIDS AND LYE.
- GRAINS PER GALLON (GPG) - A COMMON BASIS OF REPORTING WATER ANALYSIS. ONE GRAIN PER U.S. GALLON EQUALS 17.1 MILLIGRAMS PER LITER OR 17.1 PARTS PER MILLION (PPM). ONE GRAIN IS 1/7000 POUNDS OR 0.0647 GRAMS.
- HARDNESS - DISSOLVED CALCIUM AND MAGNESIUM IN WATER. COMPOUNDS OF THESE TWO ELEMENTS ARE RESPONSIBLE FOR MOST SCALING IN PIPES AND WATER HEATERS AND CAUSE NUMEROUS PROBLEMS. HARDNESS IS USUALLY EXPRESSED IN GRAINS PER GALLON AS CALCIUM CARBONATE EQUIVALENT. THE PRESENCE OF IRON IS OFTEN ASSOCIATED WITH HARDNESS.
- HARD WATER - WATER CONTAINING CALCIUM AND MAGNESIUM SALTS IN CONCENTRATION OF 1 GRAIN PER GALLON OR MORE (AS CALCIUM CARBONATE EQUIVALENT).
- IRON - AN ELEMENT OFTEN PRESENT IN GROUND WATER IN A SOLUBLE FORM (SUCH AS FERROUS BICARBONATE), USUALLY IN QUANTITIES RANGING FROM ZERO TO 10 PARTS PER MILLION. IRON IN SOLUTION IS SUSCEPTIBLE TO OXIDATION, AND WHEN CONTACTED BY AIR, PRECIPITATES AS A REDDISH-BROWN FLOC WHICH CAUSES STAINING AND DISCOLORATION.
- LIME SCALE - HARD, INSOLUBLE SCALE, CHIEFLY CALCIUM CARBONATE, COMMONLY FORMED WHEN WATER CONTAINING CALCIUM BICARBONATE IS HEATED.
- MAGNESIUM - ONE OF THE ELEMENTS MAKING UP THE EARTH'S CRUST, THE COMPOUNDS OF WHICH, WHEN DISSOLVED IN WATER, MAKE THE WATER HARD. THE PRESENCE OF MAGNESIUM IN WATER IS A FACTOR CONTRIBUTING TO THE FORMATION OF SCALE, AND INSOLUBLE SOAP CURDS.
- PARTS PER MILLION (PPM) - A COMMON BASIS OF REPORTING WATER ANALYSIS. ONE PART PER MILLION (PPM) EQUALS 1 POUND PER MILLION POUNDS OF WATER; 17.1 PPM EQUALS ONE GRAIN PER GALLON.
- pH VALUE - A TERM WHICH EXPRESSES THE BALANCE BETWEEN THE STRENGTH OF THE ALKALINE AND ACIDIC COMPOUNDS IN WATER. AT A

WATER CONDITIONING (continued)

pH OF 7.0, ACID STRENGTH AND ALKALINE STRENGTH ARE EQUAL AND THE WATER IS NEUTRAL. pH VALUES APPROACHING ZERO INDICATE INCREASING ACID STRENGTH; VALUES APPROACHING 14 INDICATE INCREASING ALKALINE STRENGTH.

- SOFT WATER - WATER CONTAINING LESS THAN 1 GRAIN PER GALLON DISSOLVED CALCIUM AND MAGNESIUM SALTS (AS CALCIUM CARBONATE EQUIVALENT).
- TURBIDITY - THIS TERM INDICATES UNDISSOLVED MINERALS IN WATER, SUCH AS FINE PARTS OF SILT, CLAY, SAND, ETC.

WINDOW PARTS-----

THE PARTS AND LOCATION OF A STANDARD DOUBLE HUNG WINDOW ARE SHOWN BELOW.

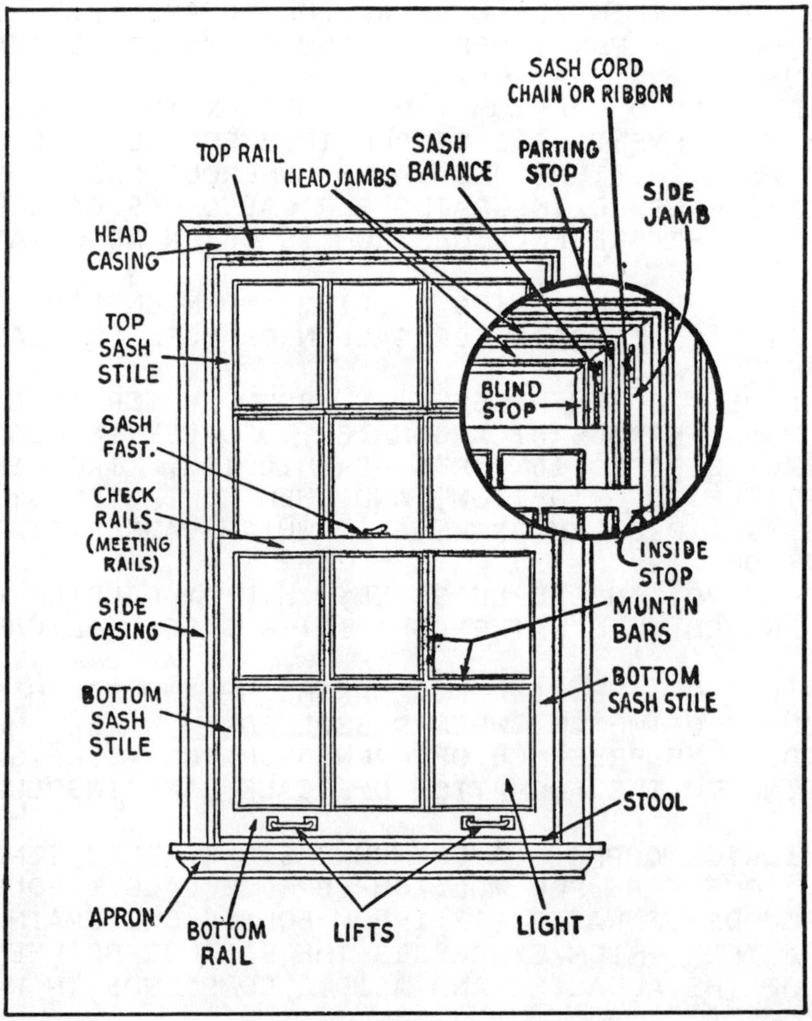

CASEMENT WINDOWS, HAND-----

THE HAND OR OPENING OF CASEMENT WINDOWS IS SHOWN HERE.

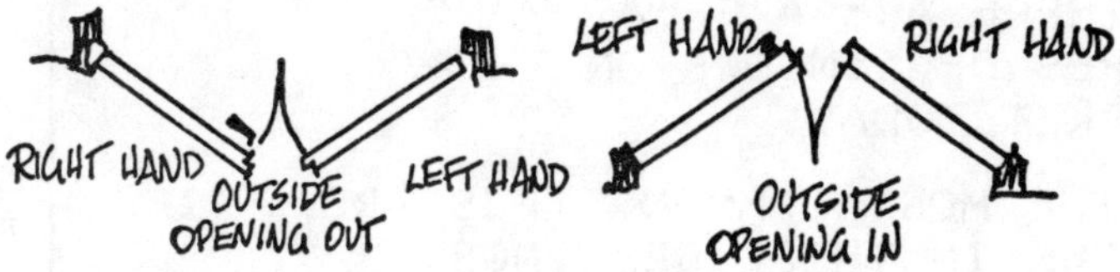

CASEMENT WINDOWS, SWING-----

THE SWING OR PROJECTION OF CASEMENTS AS YOU VIEW THEM FROM THE OUTSIDE IS INDICATED HERE.

WINDOW MEASUREMENTS-----

USE THE SKETCH BELOW TO DETERMINE PROPER MEASUREMENTS FOR WINDOW ACCESSORIES.

FOR FULL LENGTH DRAPERIES, USE THE MEASUREMENT FROM THE TOP OF THE WINDOW CASING TO THE FLOOR, LESS ONE INCH FOR CLEARANCE.

FOR CURTAINS, MEASURE FROM THE TOP OF THE WINDOW TRIM TO THE TOP OF THE SASH, TOP OF THE SILL OR BOTTOM OF THE APRON AS REQUIRED.

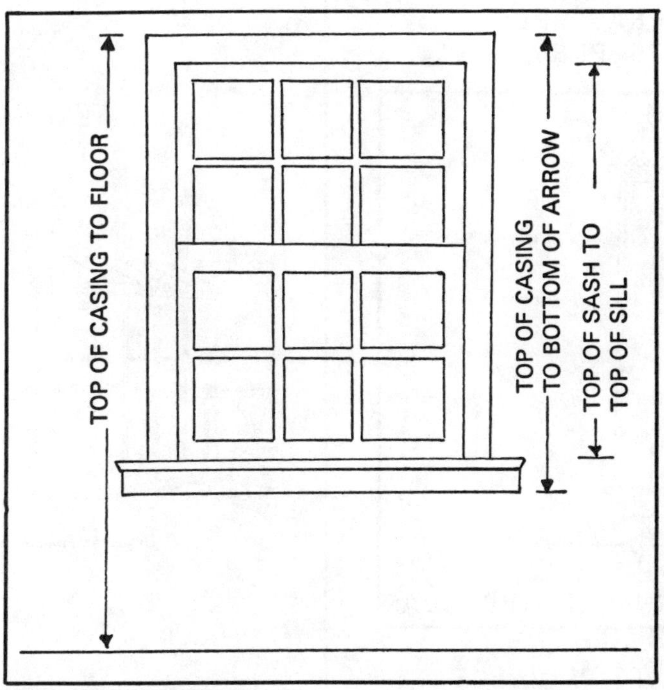

WINDOW TYPES-----

THERE ARE A GREAT MANY TYPES OF WINDOWS ON THE MARKET TODAY.

THE MOST COMMON TYPES ARE SHOWN IN THESE DRAWINGS.

OTHERS NOT SHOWN ARE: PICTURE (FIXED), JALOUSIES, PROJECTED AND BASEMENT TYPES.

MOST TYPES ARE MADE IN EITHER WOOD OR ALUMINUM. SOME ARE AVAILABLE IN VINYL AND SOME UNITS ARE BUILT OF WOOD COVERED WITH ALUMINUM OR VINYL CLADDING. THIS KIND IS GOOD FOR WEATHER AND WEAR PROTECTION AND LOW MAINTENANCE COSTS.

SEE: *WINDOW PARTS.*

WINDOW GLAZING----

DOUBLE GLAZING OF WINDOWS IS A GREAT ENERGY SAVER. HEAT LOSS CAN BE SUBSTANTIALLY REDUCED BY THIS METHOD.

GLARE PROOF, TINTED OR REFLECTIVE GLASS HELPS REDUCE COOLING COSTS IN AREAS WHERE EXTREMELY HOT SUN IS PREVALENT.

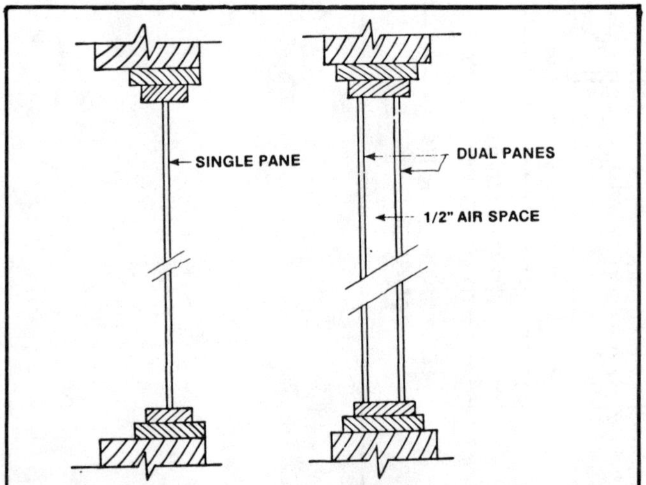

COURTESY OWENS-CORNING FIBERGLAS

SEE: *GLASS, INSULATION VALUE.*

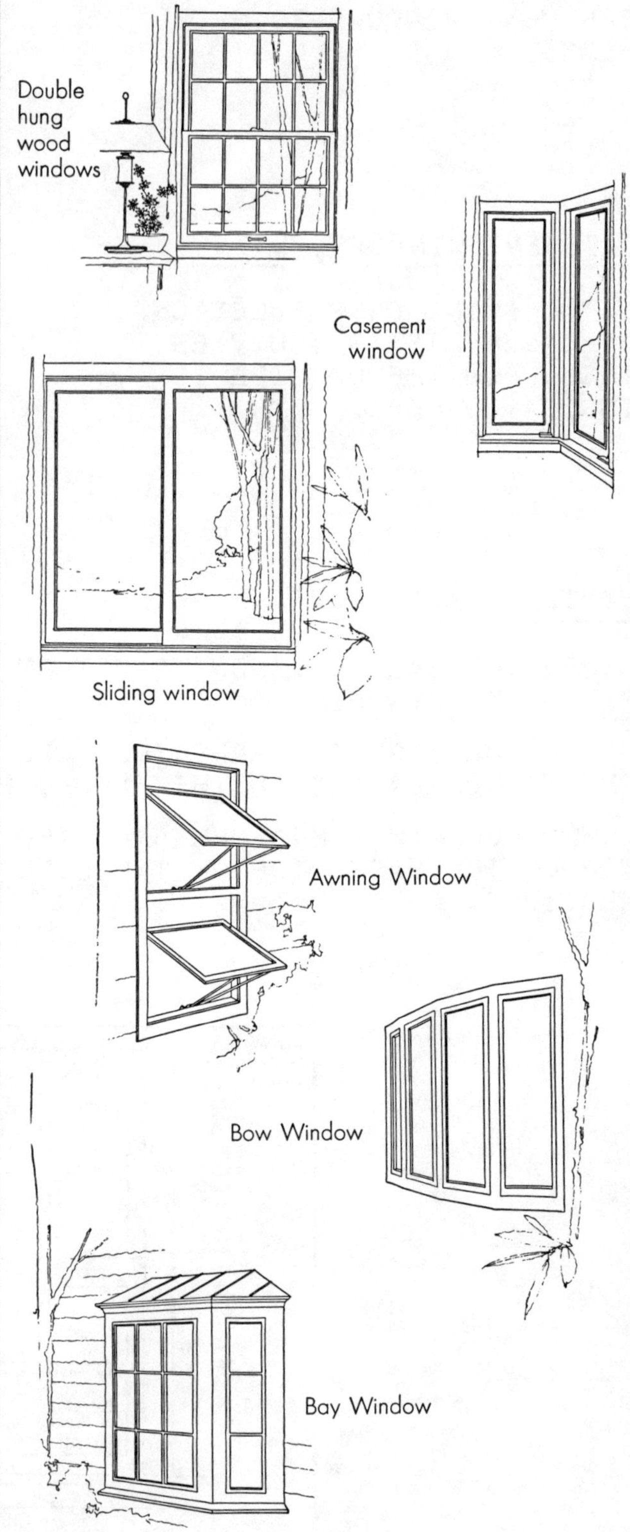

Drawings: Ponderosa Pine Woodwork

WINDOW SECTIONS-----

THE TWO CROSS-SECTIONS BELOW SHOW TWO DIFFERENT KINDS OF WINDOWS AND THEIR CONSTRUCTION DETAILS.

A. DOUBLE HUNG B. CASEMENT TYPE

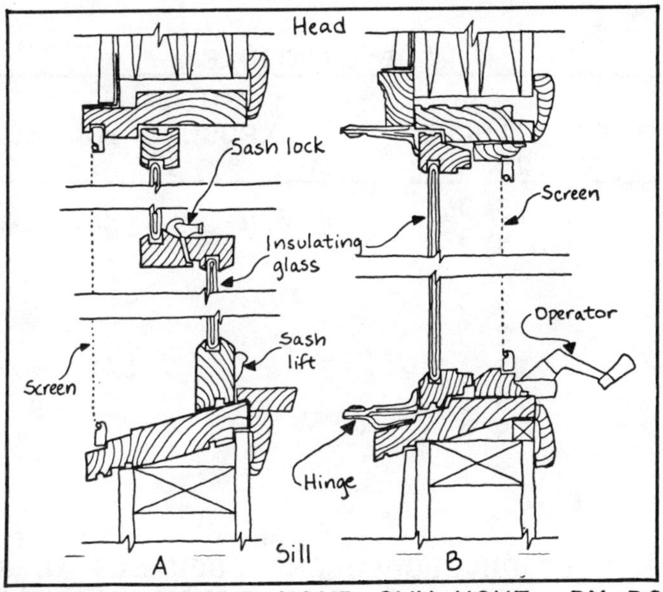

COURTESY SUCCESSFUL HOW TO BUILD YOUR OWN HOME, BY ROBERT C. RESCHKE (STRUCTURES PUBLISHING CO.)

SEE: *WINDOW PARTS*.

WOOD SHINGLES-----

WOOD SHINGLES OF VARIOUS SIZES ARE USED FOR NEW ROOFS, OVER-ROOFING (REROOFING) AND SIDE WALLS FOR HOMES AND MODERNIZATION. THEY ARE PRODUCED IN NATURAL WOOD AND IN SEVERAL PRESTAINED COLORS AND SHADES.

THREE OR FOUR GRADES ARE AVAILABLE UNSTAINED IN 16", 18" AND 24" LENGTHS. THE VARIOUS SIZES AND TYPES GIVE A GREAT VARIETY OF ROOF OR WALL PATTERNS. WOOD SHINGLES ARE SOLD BY THE SQUARE OR IN SOME CASES BY THE BUNDLE. VARIOUS TYPES ARE SHOWN IN THE SKETCH BELOW.

1. 2. 3. 4. 5. 6. 7. 8.

1. SHINGLE
2. HANDSPLIT AND RESAWED SHAKE
3. TAPERED SPLIT SHAKE
4. STRAIGHT SPLIT SHAKE
5. GROOVED WALL SHAKE
6. SANDED SHINGLE
7. KNOTTY SHINGLE
8. FANCY BUTT SHINGLE

COURTESY : WESTERN RED CEDAR SHINGLE BUREAU

WOOD SHINGLES, EXPOSURES-----

IF YOU USE WOOD SHINGLES FOR ROOFING, THIS TABLE WILL GIVE YOU THE TYPICAL EXPOSURE FOR THE 16-18-24 INCH LENGTH SHINGLES ON VARIOUS ROOF PITCHES.

ROOF EXPOSURE

PITCH	RISE INCHES TO 12" RUN	SHINGLE EXPOSURE INCHES 16"	18"	24"
1/8	3)			
1/6	4):	3-3/4	4-1/4	5-3/4
1/4	6)			
1/3	8)			
1/2	12 :	5	5-1/2	7-1/2
5/8	15)			
3/4	18)			

MANY PATTERNS MAY BE DEVELOPED ON SIDEWALLS TO SUIT THE ARCHITECTURE OR INDIVIDUAL TASTES. FOR NORMAL USE THE SUGGESTED EXPOSURE FOR 16-18-24 INCH LENGTH SHINGLES OR SHAKES IS FOUND IN THE TABLE BELOW.

SIDEWALL EXPOSURE

SHINGLE LENGTH	EXPOSURE IN INCHES SINGLE COURSE	DOUBLE COURSE
16"	6 TO 7-1/2	8 TO 12
18"	6 TO 8-1/2	9 TO 14
24"	8 TO 11	12 TO 16

WOOD-----

WOOD IS THE ONLY UNIVERSALLY USED BUILDING MATERIAL THAT IS A NATURAL AND RENEWABLE RESOURCE. THROUGH PROPER CONTROL, IT IS A MODEL ECOLOGICAL PRODUCT THAT WILL PROVIDE WORKABILITY, VERSATILITY—WITH ECONOMY AND ENDURANCE—WHEN PROPERLY INSTALLED, USED AND MAINTAINED. HERE IS A COMPREHENSIVE LIST OF MANY KINDS OF AVAILABLE WOOD. IT ALSO SHOWS THE PROPERTIES AND USES FOR MOST VARIETIES.

HARDWOODS

NAME	DESCRIPTION	HARDNESS	WORKABILITY	USES
ASH	SAPWOOD WHITE, HEARTWOOD BROWN, STRONG AND STIFF	MEDIUM	POOR	HANDLES, VEHICLE PARTS, FURNITURE, INTERIOR FINISH
BASSWOOD	SAPWOOD CREAMY WHITE, HEARTWOOD LIGHT BROWN; LIGHT IN WEIGHT; STRAIGHT GRAINED	SOFT	GOOD	MILLWORK, SIDING, KITCHEN CABINETS, BOXES, WOODENWARE, FURNITURE
BEECH	SAPWOOD WHITISH, HEARTWOOD REDDISH BROWN; HEAVY; BROAD AND NARROW WOOD RAYS OR FLECKS, EASILY VISIBLE	HARD	POOR	FURNITURE, FLOORING, INTERIOR FINISH, HANDLES, WOODENWARE, CONTAINERS
BIRCH	WHITISH SAPWOOD, LIGHT TO DARK REDDISH BROWN HEARTWOOD; FINE UNIFORMED TEXTURE; HEAVY; TAKES BEAUTIFUL NATURAL FINISH	HARD	POOR	INTERIOR FINISH, DOORS, FURNITURE, FLOORING, WOODENWARE, NOVELTIES
CHERRY	SAPWOOD NEARLY WHITE, HEARTWOOD LIGHT TO DARK REDDISH BROWN WITH BEAUTIFUL LUSTER; LOW IN SHRINKAGE	MEDIUM	POOR	FURNITURE, PANELING, FLOORING, WOODENWARE, CASKETS
GUM	SAPWOOD FLESH COLORED, HEARTWOOD OF SWEET GUM IS REDDISH BROWN, HEARTWOOD OF OTHER SPECIES LIGHT GREY; MODERATELY HEAVY; UNIFORM TEXTURE	MEDIUM	FAIR	INTERIOR FINISH, MILLWORK, CONTAINERS, FURNITURE. (ONE OF THE MOST USED FURNITURE WOODS FOR IMITATION OF WALNUT AND MAHOGANY)

(continued)

WOOD (continued)

HARDWOODS

NAME	DESCRIPTION	HARDNESS	WORK-ABILITY	USES
HICKORY	SAPWOOD WHITE, HEARTWOOD REDDISH BROWN; VERY TOUGH; HEAVY; HIGH SHOCK RESISTANCE	HARD	POOR	TOOL HANDLES, VEHICLE PARTS, SPORTING GOODS, AGRICULTURAL IMPLEMENTS, WHEELS
MAPLE	HARD MAPLE HAS WHITE SAP-WOOD AND LIGHT REDDISH BROWN HEARTWOOD; HEAVY; UNIFORM TEXTURE. (SOFT MAPLE IS LIGHTER IN COLOR, LESS HARD AND STRONG)	HARD (SOFT MAPLE, MEDIUM)	POOR	FLOORING AND OTHER MILL PRODUCTS, INTERIOR FINISH, HANDLES, ATHLETIC EQUIPMENT
OAK	SAPWOOD OFF-WHITE WITH BROWNISH HEARTWOOD IN WHITE OAK; RED OAK HEARTWOOD REDDISH BROWN; HEAVY; STRONG; PRO-NOUNCED FLECKS OR WOOD RAYS VISIBLE ON VERTICAL GRAIN LUMBER; PORES LARGE	HARD	POOR	FLOORING, INTERIOR FINISH, FURNITURE, CABINET WORK, PLANING MILL PRODUCTS; ONE OF MOST USED AMERICAN WOODS
WALNUT	NEARLY WHITE SAPWOOD, LIGHT TO DARK CHOCOLATE BROWN HEARTWOOD; HEAVY; DEVELOPS BEAUTIFUL FINISH	MEDIUM	FAIR	CABINET WORK, FURNITURE, INTERIOR FINISH, PANELING, FLOORING, GUN STOCKS
YELLOW POPLAR	SAPWOOD WHITE, HEARTWOOD YELLOWISH BROWN WITH GREENISH TINGE; LIGHT IN WEIGHT; SOFT; STRAIGHT GRAIN AND UNIFORM TEXTURE	SOFT	GOOD	SIDING, FURNITURE, PLANING MILL PRODUCTS, BOXES, INTERIOR FINISH

(continued)

SOFTWOODS

NAME	DESCRIPTION	HARDNESS	WORK-ABILITY	USES
CEDAR (NORTHERN WHITE)	HEARTWOOD LIGHT BROWN TO REDDISH TINGE; SAPWOOD LIGHT TO WHITE; LIGHT IN WEIGHT; LOW IN SHRINKAGE; RESISTANT TO DECAY; AROMATIC ODOR	SOFT	GOOD	SIDING, POSTS, POLE TIES, SHINGLES, SPECIALITY PURPOSES
CEDAR (WESTERN RED)	RED TO PINK BROWN IN COLOR EXCEPT SAPWOOD, WHICH IS WHITE; CHARACTERISTIC ODOR; HIGHLY RESISTANT TO DECAY	SOFT	GOOD	SIDING, SHINGLES, LIGHT CONSTRUCTION, PANELING, POLES, POSTS, PLANING MILL PRODUCTS, BOXES
HEMLOCK (WESTERN)	GREYLIKE WHITE WITH PURPLE TINGE; ANNUAL RINGS DISTINCT; NO RESIN DUCTS; MODERATELY LIGHT IN WEIGHT; STIFF	MEDIUM	FAIR TO GOOD	GENERAL CONSTRUCTION; PLANING MILL PRODUCTS, MILLWORK, PANELING, BOXES AND CRATES
LARCH (WESTERN)	RUSSET TO NUT BROWN HEARTWOOD, FINISHED SURFACES APPEARING SOMEWHAT WAXY; STRONG; SOMETIMES CONFUSED WITH DOUGLAS FIR	HARD	POOR	GENERAL CONSTRUCTION, PLANING MILL PRODUCTS, MILLWORK, PANELING, BOXES AND CRATES
PINE (LODGEPOLE)	SAPWOOD ALMOST WHITE, HEARTWOOD SLIGHTLY DARKER RANGING FROM CLEAR YELLOW TO PALE BROWN TINGED WITH RED; RESINOUS ODOR AND TASTE	SOFT	GOOD	SIDING, FRAMING, BOARDS, PANELING, TIES, POLES, MINING TIMBERS
PINE (PONDEROSA)	CREAMY WHITE TO LIGHT REDDISH BROWN; NARROW SUMMERWOOD RINGS; UNIFORM TEXTURE; MODERATELY STRONG	SOFT	GOOD	MILLWORK AND PLANING MILL PRODUCTS, LIGHT CONSTRUCTION, BOXES, CRATES

(continued)

SOFTWOODS

NAME	DESCRIPTION	HARDNESS	WORK-ABILITY	USES
PINE (SOUTH-ERN)	INCLUDES SEVERAL SPECIES, YELLOW TO REDDISH BROWN HEARTWOOD; RESINOUS ODOR; DISTINCT SUMMER-WOOD RINGS; MODERATELY DENSE TO DENSE; STIFF; STRONG	HARD	POOR	ALL TYPES OF BUILDING AND GENERAL CONSTRUCTION, SIDING, FLOORING, MILLWORK, DOORS, BOXES, CRATES
PINE (WHITE)	IDAHO, NORTHERN WHITE, SUGAR PINES ARE ALL GENUINE WHITE PINES; COLOR CREAMY WHITE TO LIGHT BROWN; TRANSITION FROM SPRINGWOOD TO SUMMERWOOD GRADUAL; SOFT UNIFORM TEXTURE; LIGHT; STABLE UNDER MOISTURE CHANGES	SOFT	GOOD	MILLWORK, PLANING MILL PRODUCTS, SHEATHING AND SUB-FLOORING, BOXES AND CRATES
REDWOOD	DARK, BROWN HEARTWOOD, CREAMY WHITE SAPWOOD; ODORLESS, RINGS DISTINCT; MODERATELY LIGHT AND STRONG	MEDIUM	FAIR	GENERAL CONSTRUCTION, PLANING MILL PRODUCTS, TANKS, VATS, SHINGLES, NOVELTIES
SPRUCE (SITKA)	HEARTWOOD YELLOW TO PALE BROWN WITH PURPLISH CAST, SAPWOOD CREAM TO LIGHT YELLOW IN COLOR; MODER-ATELY LIGHT IN WEIGHT; STRONG AND STIFF; HIGH STRENGTH TO WEIGHT RATIO	HARD	FAIR	PLANING MILL PRODUCTS, SID-ING, MILLWORK, FINISH, LADDERS, AIRCRAFT, BOXES AND CRATES
SPRUCE (ENGLE-MANN)	PALE OFF-WHITE COLOR; SMALL LIGHT COLORED KNOTS; FINE GRAINED; ODORLESS; LIGHT IN WEIGHT	SOFT	FAIR	PLANING MILL PRODUCTS, SID-ING, GENERAL CONSTRUCTION, FINISH, SCAFFOLDING, INDUSTRIAL USES

Other Successful Books...

SUCCESSFUL BATHROOMS, 2nd Edition, Schram. Is your bathroom as beautiful and functional as you want it to be? The smallest room in the house can be recreated easily and economically with the expert advice given here. This revised and updated edition includes water-saving devices, updated designs, remodeling, fixtures and fittings, lighting and more. Even budgeting to make the project economical. There is also new coverage of hot tubs and saunas, and a new color section. Let your creativity take over, and brightsn up this room so important to the family's daily routine. 128 pages. $6.95 Paperback.

SUCCESSFUL CHILDREN'S ROOMS, Schram. Help your child develop his uniqueness—plan and decorate specifically for him or her. This imaginative book is full of bright ideas for large or small projects in your child's world. Includes instructions for building "to-scale," "childproofing" a bathroom, making furniture, controlling noise, and providing for safety. More than 200 photos and diagrams, including color, help you visualize the results. 144 pages. $6.95 Paperback.

SUCCESSFUL CUSTOM INTERIORS, Cornell. Be your own decorator, and let your creativity flow! Here is a thorough discussion of decorating principles, with detailed plans, diagrams and instructions to make the job fun and the results professional looking. Ideas for every room in the house—even a platform tub in the bathroom! A MUST for anyone who wants to express his or her individuality in decorating. 144 pages. $6.95 Paperback.

SUCCESSFUL FAMILY AND RECREATION ROOMS, Cornell. Remodel an existing room or create your own family or recreation room from an existing room or basement! It's easy with this concise and well-illustrated book for the average handyperson. Included are all aspects of remodeling this well-used area of the home: plumbing, wiring, lighting, ventilation, fireplaces, entertainment, eating areas and MORE! Over 200 photos and illustrations guide the way to the room you've always wanted—perfect for the older home. 136 pages. $6.95 Paperback.

SUCCESSFUL FIREPLACES, 20th Edition, Lytle & Lytle. That center of warmth and sociability, the treasured fireplace, needs to be reassessed in light of today's needs. This revised edition of a best seller (over one million copies sold) covers up-to-date construction techniques, and includes material on woodburning stoves, wood as a fuel, energy savings, and safety features. Unchanging basic principles are given, as well as new ideas for making your fireplace useful and attractive. 128 pages. $6.95 Paperback.

SUCCESSFUL HOME PLANS, Pollman. Dreaming of a new home? Here's a collection of the most outstanding and popular home plans to set your imagination soaring. Included are both traditional and contemporary plans for homes of all sizes. Features family, recreation, hobby, and work rooms, storage and laundry facilities, balconies, decks and patios. Order forms are provided for complete plans and specifications that can save money in architect's fees. A special section compiled from HUD's *Manual of Acceptable Practices* will help the homeowner avoid costly mistakes. 192 pages. $5.95 Paperback.

HOMEWORK 1: Inside. Handypersons with all degree of talent will find fresh, new ideas to extend their creativity in this beautiful new volume of indoor projects. Photos and line drawings, in full color throughout, simplify the many practical designs. Includes furniture, toys, games, and accessories. Over 500 photos and illustrations. 112 pages. $6.95 Paperback.

HOMEWORK 2: Outside. Plan your outside projects for that professional look! Nearly 40 unusual outdoor designs are collected here for the home handyperson. A beautiful new volume, in full color throughout, to set your creativity to work. Includes outdoor furniture pieces, children's play equipment, garden accessories, more. Over 500 photos and illustrations. 112 pages. $6.95 Paperback.

SUCCESSFUL HOW TO BUILD YOUR OWN HOME, 2nd Edition, Reschke. A completely revised and updated edition of our bestseller. You *can* build your own home with complete and detailed information—on financing, choosing a site and plans, step-by-step construction and finishing methods. *50* chapters and over 500 photos and diagrams enable most everyone to build the home of their dreams! Passive solar heating systems, woodburning furnaces and energy considerations make this book the most up-to-date reference of its kind. 352 pages. $9.95 Paperback.

IMPROVING THE OUTSIDE OF YOUR HOME, Schram. Upgrade the value of your home and reduce maintenance, while enjoying its improved appearance. Even the novice can improve house facades and natural settings with these easy-to-follow instructions. Basic information is given on the advantages or disadvantages of various materials, and for every element from curb to chimney to rear fence. Special emphasis is given to roof, siding, windows, doors, porches, decks, landscaping and outdoor living areas. Illustrated. 152 pages. $6.95 Paperback.

SUCCESSFUL KITCHENS, 2nd Edition, Galvin. Modernize your kitchen for the best possible use of this most important room. It must be efficient and workable for today's pace! This revised edition gives thorough, reliable information on kitchen design and construction, including step-by-step instructions for such projects as counter tops and flooring. More than 200 photos and illustrations help make the process not only painless, but fun! 144 pages. $6.95 Paperback.

SUCCESSFUL LIVING ROOMS, Hedden. For only a small investment, you can add balconies, create a sunken living room, build an entertainment area or simply create a decorator look! 152 pages of ideas and detailed instructions show you how—whether you're a do-it-yourselfer or choose to hire a contractor. Over 250 color and black and white photos and illustrations. $6.95 Paperback.

OUTDOOR RECREATION PROJECTS, Bright. Plan and build your own recreation projects that cost less and fit your needs. Choose from small-scale children's play areas, hot tubs, platform tennis, or home putting greens, to larger ventures such as tennis courts, ice hockey rinks and swimming floats. Includes recommendations on materials, estimated costs, and easy-to-follow instructions. 160 pages. $6.95 Paperback.

SUCCESSFUL PAINTING, Banov & Lytle. Paint like a pro! With the advice given here and your creativity, you can transform inside or outside surfaces with very little time and money invested. Covers surface preparation, paint selection and application, problems, antiquing, refinishing, more. Non-technical language makes the job easy. For the beginner or practiced painter. 113 pages. $4.95 Paperback.

SUCCESSFUL PLAYHOUSES, Boeschen. Children want homes of their own, too! Build a playhouse with surprising ease and economy to suit their personal dreams. This gold mine of inspiration for both indoor and outdoor play will set you and your child to work on a traditional playhouse, or a treehouse, dollhouse, Mongolian Yurt or adventure garden. The physically and mentally handicapped child is considered throughout, and safety tips are emphasized. Photos and diagrams help simplify the job. 144 pages. $6.95 Paperback.

PRACTICAL AND DECORATIVE CONCRETE, Wilde. Complete information for the layman on the use of concrete inside or outside the home. The author gives instructions for the installation, maintenance, and repair of foundations, walkways, driveways, steps, embankments, fences, tree wells, patios and also suggests "fun" projects such as garden sculpture, outdoor and playground furniture; bookends. There are also chapters with easy–to–follow steps on creating fancy finishes using aggregates or colored concrete. *A selection of McGraw-Hill and Popular Science Book Clubs.* 144 pages. Paperback $6.95.

SUCCESSFUL PUTTING IT ALL TOGETHER, Scharff. Don't depend on the salesperson at the hardware store—know what you need *before* you get there! At last, a complete book of fastening devices—if you've ever wondered how to connect *anything* to anything else, this is the book for you. A must for any household! Topics include nails, screws, staples, tape, bolts, glues and adhesives, dowels, soldering, brazing, welding, electrical connectors, anchors, truss clips, rope, rivets and much more! All fastening methods are carefully detailed. Illustrated. 160 pages. $7.95 Paperback.

SUCCESSFUL HOME REPAIR, When Not to Call the Contractor, Paulsen. Anyone—from the least handy to the very handy—can cope with household repairs or emergencies using this detailed, clearly written book. Beginning with advice on how to set up a maintenance program that will help prevent the need for major repairs and take only minutes per month—the author provides tips, tricks–of–the–trade, and recommendations for dealing with over a hundred problems. Step–by–step instructions with illustrations and charts enable the householder to make all types of repairs; from foundation to roofline including plumbing, heating and utilities. 144 pages. $6.95 Paperback.

SUCCESSFUL ROOFING & SIDING, Reschke. That all–important exterior covering can make or break a home's appearance and fuel efficiency! Catch those problem areas before they become serious with this complete book that covers contractors, doing-it-yourself, materials, and budgets. Instructions also include supplemental exterior work like dormers, insulation, and gutters—all tailored for the handy or not-so-handy homeowner. 160 pages. $7.95 Paperback.

SUCCESSFUL SHELVES & BUILT-INS, Hedden. Why buy when you can build—and save! Nearly 100 projects for the home or apartment owner, with complete detailed instructions. Build shelves, counters, bookcases, cabinets, storage units, built-in furniture and more! PLUS special chapters on: Using and Joining Plywood and Particleboard, and Applying Plastic Laminate to countertops. Over 250 photos and illustrations guide the way for the beginner or skilled craftsperson. 152 pages. $6.95 Paperback.

SUCCESSFUL SPACE SAVING AT HOME, Galvin. Like magic—create space in your home or apartment where none existed! With minimal skills *anyone* can build these projects that provide more living or storage space in an existing home. Great for older homes that lack closet space. With the author's imagination, detailed instructions, and over 200 color and black and white photos, you can conjure up space under stairs, between walls, in corners, in the ceiling, and much more! 124 pages. $5.95 Paperback.

SUCCESSFUL STUDIOS AND WORK CENTERS, Davidson. Set up a convenient work center at home or office for art projects, photography, sewing, woodworking, and many other crafts. Professional or amateur, you'll find help here to put your talents to work. Covers locations, equipment, lighting, plumbing, wiring and supplies, and floor plans. Interviews with professionals tell how they handled space and work problems. Illustrated. 144 pages. $5.95 Paperback.

SUCCESSFUL SWIMMING POOLS, Derven & Nichols. Building a pool? You won't get "soaked" if you're informed. Everything the owner or would-be pool owner should know—from choosing a pool you can afford, to securing permits, purchasing accessories and building related structures. Special sections for those who wish to build it themselves, covers construction problems, safety and lighting. A *must* for anyone considering this major purchase. 250 black and white and color photos and illustrations. 152 pages. $6.95 Paperback.

WALLCOVERINGS & DECORATION, Banov, Lytle & Rossig. Be your own decorator! Let your wallcoverings express your personality and creativity. It's easy to do with the expert advice given here, starting with basic decorating principles and carrying through right up to the finished product—including color, texture, application of various wallcoverings, and artwork and lighting as finishing touches. Abundantly illustrated. 132 pages. $6.95 Paperback.

SEND FOR OUR FREE CATALOG

Available from your local bookseller or order direct from:

STRUCTURES PUBLISHING COMPANY • BOX 1002 • FARMINGTON, MICHIGAN 48024

(Please add $1.00 postage and handling for mail orders)